AI

AICHEMY

A Modern Fable That Transforms AI Confusion into Your Competitive Advantage

FORD SAEKS
Business Growth Accelerator
CSP, CPAE, MDSG

For volume book purchases, media interviews, book signings, keynote presentations or training please contact:
PROFITRICHRESULTS.COM
or call 1-316-844-0235.

Publisher's Cataloging-in-Publication (CIP) Data

Saeks, Ford, 1961–
   AI ALCHEMY: A Modern Fable that Transforms AI Confusion into Your Competitive Advantage / Ford Saeks.
      Wichita, Kansas : Prime Concepts Group Press, 2025.

First edition.   Wichita, Kansas : Prime Concepts Group Press, 2025.

   LCCN 2025919270
   ISBN 978-1-884667-49-7 (paperback)
   ISBN 978-1-884667-51-0 (hardcover)
   ISBN 978-1-884667-50-3 (Kindle/ebook)

   Subjects (LCSH): Artificial intelligence — Management. Business — Data processing. Organizational change — Management. Decision making — Data processing. Leadership.
   BISAC: BUSINESS & ECONOMICS / Leadership (BUS058000); BUSINESS & ECONOMICS / Management (BUS041000); BUSINESS & ECONOMICS / Decision-Making & Problem Solving (BUS070040); COMPUTERS / Artificial Intelligence / General (COM032000).

   Classification: LCC TBD | DDC TBD — dc23
   LC record available at https://lccn.loc.gov/2025919270

Printed in the United States of America

Brand names mentioned in this book are used for educational illustration only

PRAISE FOR AI ALCHEMY

"In the chaos of AI hype and fear, AI Alchemy provides clarity through narrative. Ford shows that AI mastery isn't about technical skills, it's about human judgment enhanced by artificial intelligence. The fable format makes complex principles simple and memorable."

— Scott McKain, Founder and CEO, The Distinction Institute

"Most AI books tell you what to do but not how to think. Ford's approach is different. This book teaches you to see patterns, ask better questions, and make decisions that actually stick. The fable format made easy to follow and fun to read."

— Joe Mohay, President of Franchise
& MultiLocation, Ignite Visibility.

"We spent two years chasing AI solutions that solved the wrong problems. Ford's frameworks helped us step back and focus on what really matters: using AI to amplify human strengths, not replace them. The clarity we gained from AI Alchemy transformed our entire approach."

— Steve White, President & COO, PuroClean USA

"I've seen too many leaders get swept up in AI promises or paralyzed by AI fears. Ford shows a third path: measured confidence backed by proven principles. AI Alchemy gave me the language to help my clients navigate AI transformation without losing their humanity."

— Guy Coffey, Host, The Franchise Scale-Up Show

"Working with executives across industries, I see the same AI adoption mistakes repeatedly. Ford's storytelling approach breaks through resistance and creates genuine understanding."

— Patricia Fripp, President, Fripp & Associates

"Ford understands something crucial: successful AI adoption is about leadership, not technology. AI Alchemy provides the framework I recommend to executives who want to drive real change. The story format makes complex strategy accessible and memorable for entire leadership teams and his keynote audiences."
— Katrina Mitchell, CEO, Franchise Speakers

"After years of helping businesses optimize operations, I recognize Ford's wisdom immediately. AI Alchemy cuts through the noise and focuses on what drives real results: smart implementation guided by human judgment. This book should be required reading for any serious business leader."
— Joe Prenatt, Founder, Pressure Washer Sales & Service LLC

"I recognized myself in Alex's AI struggles immediately. Ford's fable format made the solution clear, it's about purpose over technology, wisdom over speed. AI Alchemy gave me the confidence to lead AI transformation instead of just reacting to it."
— Art Leger, CEO, Clean Rite Pressure Washing

"I was paralyzed by AI complexity. AI Alchemy showed me how to find the middle ground where real AI success happens. The story format makes the principles stick in ways technical manuals never could."
— Craig Merrills, Franchise Partner, WOW 1 DAY PAINTING

"Having implemented AI across three companies, I've faced many of the challenges Ford highlights firsthand. AI Alchemy distills those lessons into clear, actionable frameworks that could have saved me millions in trial and error. The principles Alex learns are the same ones that drive real results in forward-thinking organizations."
— Daniel Burrus, New York Times Bestselling Author

"Alex's journey from AI confusion to mastery mirrors what I see in every organization. Ford's frameworks cut through the hype and show exactly how to build sustainable AI advantages without losing what makes businesses human. Our entire leadership team has adopted his approach."

— Uri Geva, CEO, Cookie Dough Bliss Franchisor

"As someone building cutting-edge health technology, I thought I had AI figured out until I read AI Alchemy. Ford's fable revealed the gaps in my thinking and showed how to balance innovation with human wisdom. The story format makes complex AI strategy surprisingly accessible."

— Cody Beard, Franchisee, Prime IV Wichita | Medical Spa

"AI Alchemy captures the real challenge of AI adoption, it's not technical, it's human. Ford's story of transformation resonates because it mirrors the journey every leader faces. The frameworks are practical, but the storytelling is what makes them powerful."

— Shep Hyken, CX Expert and NYT Bestselling Author

"After watching our AI initiative stall for six months, I read AI Alchemy in one sitting. Ford's approach helped me identify exactly where we went wrong and gave our team the framework to restart with confidence. The wisdom in this fable bridges the gap between AI theory and real-world success."

— Tim Gard, CSP, Hall of Fame Keynote Speaker

"After reading dozens of AI books, AI Alchemy stands apart. Ford's fable approach makes timeless principles memorable while showing exactly how to avoid the overconfidence trap that destroys so many AI implementations. Essential reading for strategic leaders."

— Will Ezell, CEO, Biz Visioneers

Your Journey

Unlock Exclusive AI Tools & Resources

Congratulations on taking the next step to accelerate your results with AI! To help you get even more value from your purchase of *AI ALCHEMY*, I'm offering **exclusive access** to my AI Book Toolkit, designed to provide you with the essential tools, templates, and strategies to implement AI in your business right away.

Here's What You'll Get:

Bonus Video Tutorials: Learn directly from Ford Saeks with actionable tips for maximizing your AI efforts.

AI Prompt Templates: Ready-to-use prompts that can generate high-quality content, marketing strategies, and sales solutions.

Step-by-Step Action Plans: Follow easy-to-implement AI strategies that will help you streamline tasks and boost productivity.

Early Access to AI Webinars: Be the first to know about upcoming live training sessions and interactive webinars.

Scan the QR code or simply visit: **ProfitRichResults.com/aibooktoolkit** to unlock these powerful resources and start applying AI to your business today!

Author's Note

A Personal Message from Me, Ford Saeks

Before we begin, I want to be transparent with you: the story you're about to read is fiction—but the principles and frameworks it illustrates are very real."

The characters, including Alex Morgan and Maya Chen, as well as the company scenarios depicted in this book, are entirely fictional. They were created to illustrate real principles I've developed over three decades of helping leaders navigate technological transformation. While Alex's journey from AI confusion to confident mastery reflects the collective experiences of many executives I've worked with, any resemblance to actual persons, living or dead, is purely coincidental.

The frameworks, methodologies, and insights shared throughout this book are absolutely real. They're based on my actual work with organizations across every industry, from Fortune 500 corporations to franchise systems to professional associations. I've refined these approaches through thousands of hours of consulting, keynote presentations, and hands-on AI implementation projects.

Stories bypass resistance. They let us experience transformation before we try to lead it. When I watch leaders struggle with AI adoption, they all face similar challenges: overwhelming complexity, fear of replacement, confusion about where to start, and pressure to keep up with rapidly evolving capabilities. Alex's story represents this universal experience while demonstrating proven pathways to success.

Important Disclaimers

About AI References: The AI landscape evolves incredibly rapidly. Specific AI models, capabilities, and platform names mentioned in this book are used for illustrative purposes. Some references may describe emerging or hypothetical

AI capabilities to demonstrate how these principles apply regardless of technological advancement. Always verify current capabilities and limitations before making business decisions.

About Company and Product References:

Any mention of specific companies, products, or services is for narrative and educational purposes only. These references don't constitute endorsements, and any resemblance to actual business situations is coincidental.

About the Frameworks:

While the story is fictional, the business frameworks and AI implementation strategies are based on my real-world experience. However, every organization is unique. These principles should be adapted to your specific situation, industry, and risk tolerance. Consider consulting with qualified professionals before making significant business or technology decisions.

Why This Matters to You

I've spent three decades watching leaders navigate transformational change. Companies thrive by embracing new technologies thoughtfully, while others fail by rushing headlong into change or being paralyzed by fear.

AI represents the most significant technological shift in our lifetime. Leaders who master human-AI collaboration will build sustainable competitive advantages. Those who don't will find themselves replaced not by AI, but by competitors who use AI more effectively.

This book gives you the frameworks I use with my clients to transform AI overwhelm into competitive advantage. The story format makes these concepts memorable and actionable, grounded in real business experience.

Your journey from AI confusion to confident mastery can follow Alex's path. The frameworks are proven. The principles are timeless. The opportunity is yours. Please enjoy the story...

Introduction: The Quest Begins

Finding Clarity in AI Overwhelm

"The cave you fear to enter holds the treasure you seek."
— Ancient Proverb

Three meetings. Three different AI consultants.

Three promises that this time, "AI changes everything."

Alex Morgan pushed back from the conference table, looked around at the eager faces of yet another AI implementation team, and thought: *Here we go again.*

The consultant, barely old enough to rent a car, drew circles on the whiteboard with the energy of someone who had never actually run a business. "So once we integrate the comprehensive AI ecosystem with multimodal capabilities, autonomous agents, and workflow automation, you'll see transformational productivity gains."

Alex had heard variations of this pitch six times in the past year. Each consultant promised a different vision of the AI-powered future: chatbots that could see and analyze store photos, AI agents that could browse the web and execute complex workflows, voice assistants that could manage entire business processes, automated systems connecting everything from customer service to inventory management.

They all had one thing in common. None had actually solved the problem Alex was facing. Each presentation sparkled with jargon and dashboards, but none answered the real questions: How do you make sense of AI fast enough to lead your people through it? How do you make sense of AI when it's evolving faster than anyone can keep up with?

Look, I've been helping business leaders navigate transformational change for over 30 years. I've watched technology waves come and go. I've seen companies get swept away by the latest capabilities, and I've seen others miss

3

breakthrough opportunities because they waited too long to act.

AI is different. It is not just another software upgrade or process improvement. It is a full-scale business reinvention that reaches into every corner of leadership, culture, and customer experience.

And that is exactly why so many executives hesitate.

They are not afraid of technology. They are afraid of what comes with it.

They ask tough questions like these:

- What happens to our people if AI automates part of their work?
- How do we handle privacy, intellectual property, and liability when AI makes mistakes?
- What will it cost us if we get this wrong financially, legally, or reputationally?

And if our marketing is not optimized for AI discovery through AEO and structured data, how much business are we losing because we are invisible in Google's rich snippets and AI-generated search results?

These are the quiet fears that never make it into board minutes but keep decision-makers awake at night.

In just two years, AI has evolved from simple text generation to systems that can see, hear, speak, analyze images, browse the web in real time, and execute complex multi-step workflows autonomously.

Your team might already be experimenting with these tools. Some use ChatGPT for writing, Copilot for analysis, Claude for research, or Perplexity for web tasks. Others rely on voice assistants for scheduling or workflow platforms like Zapier to connect everything together.

Every tool promises transformation. Every vendor insists that this one will change everything. Yet each comes with new dependencies, risks, and learning curves. And few people can explain how to bring it all together in a way that is strategic, ethical, and profitable.

Here is what I have discovered working with organizations across every industry. The secret is not mastering every new AI capability as it emerges. The secret is understanding the timeless leadership principles that make any AI initiative— whether text, voice, vision, or automation—more effective, more trustworthy, and more human.

That is what this book, and Alex's story, will show you.

Why I'm Telling You This Story

I could have written another "Master 50 AI Tools" guide. Lord knows there are enough of those already, and they're obsolete before they're published. Instead, I decided to tell you Alex's story because stories reveal truths that technology tutorials can't capture.

Alex represents thousands of business leaders I've worked with who are smart, experienced professionals who don't need more information about AI capabilities. They need wisdom about how to think when the tools keep changing faster than anyone can learn them.

This isn't fiction. Alex's journey is a composite of real challenges, failures, and breakthroughs I've witnessed while helping organizations navigate AI transformation across every conceivable application, from simple chatbots to complex autonomous workflows.

Through my work as a keynote speaker, business strategist, and author of *AI Mindshift*, *Accelerate*, and *Superpower*, I've learned that successful AI adoption has nothing to do with technical mastery and everything to do with understanding the human side of artificial intelligence.

The Promise I'm Making You

By the time you finish Alex's story, you'll understand something most people miss: AI mastery isn't about keeping up with every new capability. It's about understanding the universal principles that make any AI interaction, current or future, more successful. Real success won't come from the next AI update, but from mastering how you think and communicate when everything around you keeps changing.

You'll discover why the biggest AI frustrations aren't about the technology. They're about communication, judgment, and maintaining human wisdom in an artificially intelligent world.

Most importantly, you'll learn how to lead AI adoption without losing the trust of your team, the loyalty of your customers, or your own sanity as the capabilities continue to evolve at breakneck speed.

What Makes This Different

This isn't a book about specific AI tools or techniques. It's about the principles that transcend any individual technology:

- **Purpose over Platform**: Why your goals matter more than which AI system you use

- **Communication over Complexity**: How to get better results from any AI through better conversation

- **Wisdom over Speed**: When to trust AI and when human judgment is irreplaceable

- **Human Connection over Artificial Intelligence**: Why the most powerful AI implementations amplify rather than replace human expertise

Whether you're using today's ChatGPT, tomorrow's autonomous agents, or whatever gets invented next year, these principles will make you more effective.

Here's How This Works

This book is built in three parts, like any good transformation.

Each part builds on the one before it, just like real transformation. Read them in sequence and you'll move from confusion to clarity, and from ideas to confident action

Part I: The Fable

Alex's journey from AI confusion to confident mastery. Don't skim this thinking you'll get to the "real stuff" later. The story IS the real stuff. Pay attention to what resonates, what challenges feel familiar, and what breakthroughs emerge. Alex's struggles with rapidly evolving AI capabilities mirror what every leader faces today.

Part II: The Framework

The practical principles behind the transformation. These aren't tips for specific AI tools that become obsolete when new capabilities launch. These are universal truths that work whether you're chatting with AI, analyzing images, managing workflows, or collaborating with autonomous agents.

Part III: Your Application

How to become an AI alchemist in your specific role. Whether you're running a small team or a Fortune 500 company, you'll find concrete ways to apply these insights across any combination of AI capabilities.

The Future Is Already Here

Alex's journey doesn't begin in a research lab or a tech conference. It begins in a place you'll recognize: a regular company with regular problems, run by regular people under pressure to figure out this rapidly evolving AI landscape.

The transformation that follows isn't about predicting the future of AI. It's about developing the timeless skills that let you adapt confidently no matter how the technology evolves.

Understanding the AI Tools Mentioned in This Book

Throughout this story, you'll see references to several popular large language models (LLMs)—the advanced conversational AIs that help people write, analyze, brainstorm, and automate work. You don't need to be technical to follow along.

Here are the main ones you'll see mentioned, with direct links if you want to explore them:

ChatGPT (OpenAI) – https://chat.openai.com

The most widely used conversational AI. It can generate text, summarize, analyze data, and now handle images, voice, and web browsing.

Claude (Anthropic) – https://claude.ai

Known for its friendly tone and safety-first design, Claude excels at reasoning, summarizing long documents, and maintaining context across conversations.

Gemini (Google DeepMind) – https://gemini.google.com

Google's AI assistant that connects deeply with Gmail, Docs, Sheets, and Search—great for those already using Google Workspace.

Perplexity – https://www.perplexity.ai

A conversational search engine that combines AI reasoning with real-time, cited web results. Think of it as AI-powered research with sources you can verify.

Copilot (Microsoft) – https://copilot.microsoft.com

Integrated into Microsoft 365 (Word, Excel, Outlook), Copilot helps automate writing, analysis, and presentations directly inside familiar apps.

These examples will help you understand the references in Alex's story. They're all versions of the same larger concept—AI systems that can communicate, analyze, and create alongside humans. The specific tools may evolve, but the universal principles you'll learn in these pages apply to all of them.

PART I:
THE FABLE OF AI
ALCHEMY

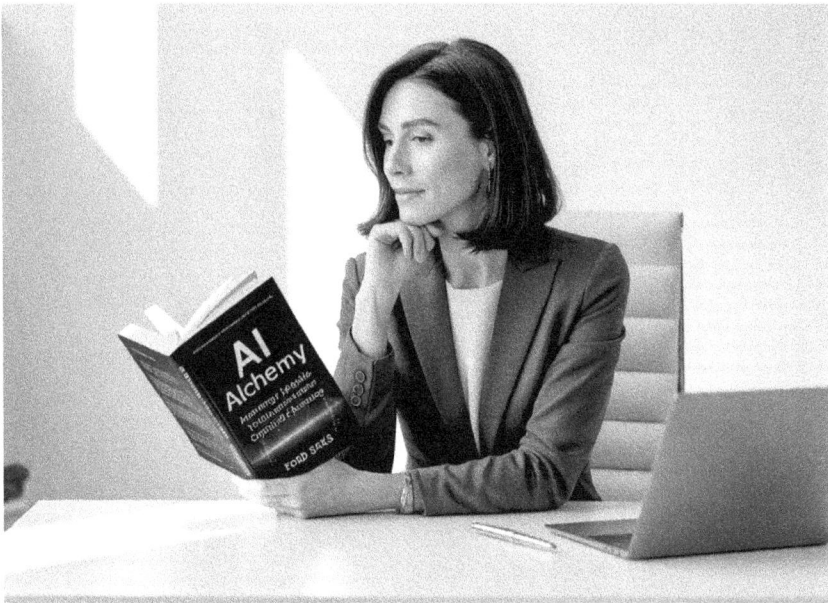

Chapter 1: The Leader and the AI Oracle

When Expert Advice Leads You Nowhere

Alex Morgan stared at the laptop screen, cursor blinking mockingly in the empty ChatGPT window.

As Director of Franchise Operations for a sixty-location retail chain, Alex had seen plenty of new technologies come and go. But this AI wave felt different, more urgent, more confusing, and definitely more hyped than anything before.

Alex thought about the conversation with Sarah Chen during the hiring interview eighteen months ago. 'You'll be our bridge between corporate innovation and franchise reality,' she'd said. 'That's why we're betting on you.' Betting. The word still stung.

At forty-two, after three companies and a career restart that had already cost two years of prime earning time, this job was supposed to be the one that stuck. The one that proved the pivot wasn't a mistake. The one that finally showed his teenage kids their father hadn't given up, just adapted. Now, staring at this blinking cursor, that bet felt increasingly risky.

Everyone else made this look so easy.

Just last week, Alex had watched a colleague generate a perfect presentation in five minutes using Copilot. The marketing intern was using AI to analyze customer photos and create targeted social campaigns. Even Alex's teenager was having voice conversations with Perplexity.ai for homework research, getting better grades than ever.

Meanwhile, Alex's attempts at AI usually ended the same way: confusing results, wasted time, and the growing suspicion that everyone else had received some secret instruction manual for this rapidly evolving AI world.

"You just need to learn the fundamentals," said Jake from IT, with the confidence of someone who'd watched three YouTube videos. "Whether you're using text, voice, images, or workflows, it's all about understanding how these AI systems process what you give them."

Understanding how AI systems think. Sure.

Alex had tried everything. Copied prompts from online articles. Followed step-by-step tutorials for different AI capabilities. Even bought a $97 course called "*Master Every AI Tool: From Chatbots to Autonomous Agents.*" The results ranged from mediocre to completely useless, regardless of which AI capability was being used.

But the worst part wasn't the bad results. The worst part was watching everyone else act like they had it all figured out.

LinkedIn was flooded with posts about "*10X productivity with multimodal AI*" and "*How I built autonomous workflows that run my business.*" Colleagues casually mentioned using Copilot for spreadsheet analysis, Claude for writing, Perplexity for web research, voice assistants for complex scheduling, and AI workflow tools that somehow connected everything together seamlessly. The company had even hired Dr. Marcus Wellington, a certified AI expert, to help everyone "leverage the full spectrum of artificial intelligence capabilities across all platforms and modalities."

"The fundamental issue," Dr. Wellington had explained in the all-hands meeting, "is that most people approach AI tools through legacy thinking patterns. Whether you're using large language models like ChatGPT and Claude, multimodal systems that can see and hear, autonomous web-browsing agents, or workflow automation platforms, you need to embrace cognitive optimization across the entire AI ecosystem."

Alex had raised a hand. "Can you just show us how to get better results from the tools we're already using?"

Dr. Wellington smiled like a professor addressing a kindergartner. "First, you need to understand the underlying architecture differences between text-based AI, computer vision systems, autonomous agent frameworks, and workflow orchestration platforms. Each modality requires different optimization strategies."

Forty-five minutes later, Alex was more confused than ever. The presentation wandered through machine learning architectures, multimodal integrations, and something called agentic workflow optimization across distributed AI systems."

Not one example of actually getting better results from any specific capability.

That evening, Alex called Chris, a friend who ran a small consulting business.

"Please tell me you've figured out this AI thing," Alex said.

"Honestly? I feel like I'm back in math class, pretending I understand while everyone else nods along."

It was the first honest thing Alex had heard all week

"But your LinkedIn posts make it sound like AI transformed your business."

"That's the weird part. It actually has helped. But not because I understand the technical differences between all these AI capabilities. I just keep trying different tools for different tasks until something works."

Alex felt a wave of relief. "So you don't know the difference between multimodal AI and autonomous agents either?"

"I know some AI can look at images now, and some can browse the web for me. I use voice mode in ChatGPT when I'm driving. But mostly I just ask these tools to help me with regular business stuff and sometimes it works great, sometimes it's garbage. I have no idea why."

After hanging up, Alex realized something important. The problem wasn't a lack of intelligence or technical ability. The problem was that everyone was pretending AI was either incredibly simple ("just use these 20 different AI capabilities!") or incredibly complex ("you need to understand multimodal architectures and agentic frameworks").

Both answers felt wrong.

AI clearly wasn't simple, or Alex would already have it figured out. But it also couldn't be that complex, or regular people wouldn't be using voice assistants, image analysis tools, and automated workflows successfully every day.

What if the real issue wasn't understanding each AI capability better? What if it was understanding how to think about all AI tools differently?

What if instead of trying to become an expert in text AI, voice AI, image AI, web-browsing agents, and workflow automation, Alex needed to find someone who could translate the entire AI landscape into universal principles that worked regardless of the specific capability?

Someone who understood that most people didn't want to build AI systems or master dozens of different modalities. They just wanted to use AI to do their regular work better, regardless of whether they were typing, talking, sharing images, or automating workflows.

The next morning, Alex started looking for a different kind of help entirely.

The Alchemist's Insight

AI confusion isn't about intelligence, it's about universal principles. The gap between AI capability and practical results isn't about mastering every new feature or modality. It's about understanding the fundamental communication patterns that work across all AI interactions.

You don't need to become an expert in every AI capability. You need to find someone who can teach you the universal principles that make any AI interaction more effective.

Your Next Step

Use AI to solve AI questions. Write down three tasks that frustrate you most in your daily work. The right AI approach will show you simple ways to address these challenges using whatever AI capabilities work best for your situation.

Look for universal principles, not capability-specific tactics.

Chapter 2: The Prompting Paradox

Why AI Never Gives You What You Actually Asked For

Alex typed, "Write me a professional email," into ChatGPT.

The response came back instantly: a perfectly formatted, grammatically correct email that sounded like it was written by a robot having a bad day at corporate headquarters.

Alex deleted it and tried Microsoft Copilot instead. Same request, same robotic result.

Maybe voice mode in ChatGPT would work better. "Help me write a professional email about a project delay."

"I found several articles about writing professional emails. Would you like me to read them to you?"

Not helpful.

Alex even tried uploading a photo of the handwritten notes about the project to ChatGPT's vision feature, hoping visual context would help. The AI described the notes accurately but still produced the same generic corporate-speak email.

Alex had been at this for two hours, bouncing between ChatGPT, Copilot, Claude, voice assistants, and even trying Perplexity's web browsing to research email templates. Every AI capability—text, voice, image analysis, web search—gave technically correct but completely useless results.

After two hours of frustration, Alex scrolled through Slack Jessica from accounting had just posted: "Between Copilot in Excel, ChatGPT for emails, voice prompts for scheduling, and using AI to analyze our customer service photos, I'm getting more done before lunch than I used to all day! These tools are amazing! 🚀"

What was Jessica doing that Alex wasn't?

Alex had tried everything. Different platforms, different modalities, even AI workflow tools that promised to automate entire processes. The pattern was always the same: brilliant capabilities, disappointing results.

Last week, Alex had asked Claude to summarize a market research report. Great summary, wrong focus. Asked Copilot to analyze spreadsheet data. Perfect charts, missed the insights. Even tried having AI analyze photos of competitor store layouts for strategic insights. Accurate descriptions, useless conclusions.

It was like having access to AI that could see, hear, speak, write, and browse the web, but somehow all of these capabilities managed to answer slightly different questions than the ones you asked.

Alex decided to watch Jessica work. She was orchestrating AI capabilities like a conductor with a full symphony, but there was something methodical about her approach that Alex had missed before.

"Here's what I actually do," Jessica explained, pulling up Perplexity on her screen. "I don't start by asking ChatGPT or Claude to solve my problem. I start by asking Perplexity to help me figure out what questions I should be asking in the first place."

She demonstrated with a vendor payment issue: "Help me list the key questions I should ask before writing a response to a vendor demanding payment even though delivery was late. What context and expectations should I provide to get the most useful response from AI tools?"

Perplexity responded with a structured breakdown of considerations: relationship dynamics, contract terms, communication tone, business objectives, and industry standards.

"Now I have a framework for thinking through the problem," Jessica continued. "I take these insights and craft a comprehensive prompt that I can use across different AI platforms."

For email using voice: "I need to respond to a vendor who delivered materials late and is now demanding full payment despite our contract terms. I'm the accounting manager at a mid-sized construction company. The vendor is someone we've worked with for three years and want to keep, but they need to understand we have policies. I want to be firm about our payment terms but maintain the relationship. Make it sound professional but not corporate, more like a straight conversation between business people who respect each other."

"But here's the key," Jessica said, "I don't stop with one AI tool. I take this same refined prompt to ChatGPT and Claude to see different perspectives, then bring the best elements together."

For visual analysis with ChatGPT: She uploaded a photo of their warehouse layout and asked, "Analyze this warehouse setup and suggest improvements for efficiency. Focus on areas where we're losing time in our picking process. Present recommendations that I can share with our operations team who know this space well but might not see obvious optimization opportunities."

"And here's my secret weapon," Jessica added. "After I get the initial response, I always ask: 'What additional questions should I be asking that I haven't considered? What other information would help you provide more accurate recommendations?'"

The AI responded with follow-up questions about seasonal variations, staff experience levels, budget constraints, and regulatory requirements that Jessica hadn't initially thought to mention.

"This turns it into a real conversation," Jessica explained. "I'm not just getting one answer, I'm collaborating to uncover better questions and more comprehensive solutions."

Each AI capability; text, voice, vision, web browsing, automation, gave Jessica exactly what she needed because she told them exactly what she was trying to accomplish, regardless of the modality.

"The secret," Jessica concluded, "is that I don't treat any AI capability like Google. Whether I'm typing into ChatGPT, talking in voice mode, uploading images for analysis, or setting up automated workflows, I treat them all like I'm hiring expert consultants who need really clear instructions."

Alex realized Jessica wasn't using AI faster; she was thinking with AI differently. He tried Jessica's approach across different AI modalities:

Starting with Perplexity: "Help me craft effective prompts for communicating a project delay to a client. What context and expectations should I provide to get professional, reassuring responses from AI tools?"

Then with ChatGPT for writing: "I'm a project manager writing to a long-term client. We've worked together for two years, and he appreciates direct communication. Our deliverable is delayed one week due to a vendor issue that's now resolved. I need to tell him about the delay while reassuring him we're handling it well. Make it sound confident and proactive, like someone who has everything under control even when problems come up."

After the initial response, Alex asked: "What additional questions should I be asking to make this communication more effective?"

With voice prompts while reviewing project data: "I'm preparing for a client meeting about project delays. Help me identify which tasks are behind schedule and the root causes. I need to present solutions, not just problems. Focus on what we're doing to prevent future delays."

With image analysis: Alex uploaded a photo of the project timeline chart and asked, "Look at this project timeline and identify the bottlenecks that are causing delays. Help me explain these visually to my client in a way that shows we understand the problems and have solutions."

Every AI capability gave dramatically better results when Alex provided context and expectations, regardless of whether the interaction was through text, voice, image, or web research. But the cross-platform validation Jessica had shown made the insights even richer.

But Alex noticed something else. Even with better inputs across all these different AI modalities, every response needed human judgment. ChatGPT wrote a great email but promised faster delivery than the team could handle. Voice analysis identified the right patterns but suggested solutions that ignored budget constraints. Image analysis spotted layout issues but recommended changes that weren't practical given space limitations.

Alex started treating each AI interaction; whether text, voice, image, or web-based the same way: use AI for the heavy lifting, then apply human oversight. Let it draft, then edit. Let it analyze, then verify. Let it suggest, then decide.

That evening, Alex called Trent, a freelancer who seemed to effortlessly use every AI capability available.

"I'm getting better results across different AI modalities," Alex said, "but I feel like I have to babysit everything."

Trent laughed. "That's exactly right. Whether you're using ChatGPT, voice assistants, image analysis, web-browsing agents, or workflow automation, the principle is the same: treat each capability like hiring the smartest intern you've ever met. Brilliant at execution, terrible at judgment about your specific situation."

"So you double-check everything, regardless of which AI capability you're using?"

"Everything important, yeah. Doesn't matter if it's text generation, image analysis, voice interaction, or autonomous web research, AI can give you perfect-sounding advice that's completely wrong for your situation. The capability might be different, but the need for human oversight is constant."

That made sense. The breakthrough wasn't about finding the right AI capability or modality. It was about understanding the relationship with all of them.

Context. Expectations. Conversation. Human oversight
Four principles that worked whether you were typing, talking, sharing images, or setting up workflows all enhanced by starting with better questions and validating across platforms.

The Alchemist's Insight

The magic isn't in the AI tool, it's in the quality of your questions.

Transform generic AI responses into strategic insights by starting with better questions, having cross-platform conversations, and always asking AI what you haven't considered yet. .

Whether you're using text, voice, image analysis, web browsing, or workflow automation, the **CECH** approach works:

Set **CONTEXT** (what's the situation?),

Define **EXPECTATIONS** (what does success look like?),

Engage in **CONVERSATION** (refine and iterate),

And maintain **HUMAN** oversight (you're the final decision maker).

Every AI capability is brilliant at execution but needs human judgment. The modality changes, the principle doesn't.

Your Next Step

Review the prompting checklist, prompt guide, and example prompts at the back of this book. Pick one current challenge and apply the Question-First Method:

Use Perplexity.ai to explore "What are the key questions I should ask about [your challenge]?" Take those refined prompts to two different AI tools (ChatGPT, Claude, or Copilot) to gather diverse perspectives. Then ask your follow-up power question: "What additional questions should I be asking that I haven't considered?" This transforms simple requests into strategic thinking sessions that compound in value.

Bonus Video Training:

Unleashing AI Innovations, What You Need to Know Now (constantly updated) Get instant access to my On-demand AI Masterclass video training that expand on these techniques

visit: **ProfitRichResults.com/ai-training**

Chapter 3: The Trap of Overconfidence

How Your Early Wins Set You Up for Spectacular Failure

Three weeks after applying the CECH approach, Alex felt unstoppable.

Sarah Chen, the CEO who'd hired Alex eighteen months ago to modernize their sixty-location franchise network, had been impressed with the recent efficiency gains. "Your team's turnaround time is up forty percent," she told him during their weekly one-on-one. "Whatever you're doing with these AI tools, keep it up."

That validation felt good. Really good. Maybe too good.

The Morrison Project was Alex's chance to prove the AI approach could work on high-stakes client consulting, not just internal operations. Morrison Food Services, a regional chain with ambitions to scale nationally, represented exactly the kind of sophisticated client their firm wanted to attract. Landing this account could mean a promotion. Losing it could mean... well, Alex didn't want to think about that.

ChatGPT was cranking out decent emails. Copilot was making spreadsheet analysis faster. Even voice commands to Claude were producing useful research summaries. Alex had gone from AI skeptic to AI success story in record time. And like most early successes, it built more confidence than caution.

"I think I've figured this out," Alex told Sarah during their weekly check-in. "It's all about giving these tools better context and clear expectations. Once you crack that code, AI becomes incredibly powerful."

Sarah smiled. "That's great. How's the Morrison project coming along? By the way, we just hired a new data strategist who's supposed to help us all think more strategically about AI implementation. You'll meet her at the franchise conference next month."

"Actually, that's a perfect example," Alex said, pulling up a document. "I used AI to help analyze their market data, write the proposal, and even create talking points for tomorrow's presentation. Saved me probably 20 hours of work."

What Alex didn't mention was the growing confidence that AI could handle almost anything. Why spend hours researching when Perplexity could do it in minutes? Why labor over presentations when Copilot could generate slides instantly? Why struggle with complex analysis when ChatGPT could break down any problem?

The Morrison presentation was scheduled for 2 PM the next day.

At 1:45 PM, Alex was feeling confident. The AI-generated presentation was polished, comprehensive, and impressive. The talking points were sharp. The market analysis looked thorough.

"So," Morrison said after Alex finished the pitch, "your data shows our primary competitor is increasing market share by 15% annually?"

"That's correct," Alex replied, referencing the AI-generated analysis.

Morrison frowned. "Interesting. Because they filed for bankruptcy protection last month."

Alex's stomach dropped. This wasn't just embarrassing, Morrison was the kind of client who could make or break reputations in the industry. Sarah had specifically asked Alex to handle this personally because of the recent AI success with internal projects.

"And this recommendation to expand into the automotive sector," Morrison continued, "you realize we're a food service company, right? We don't have automotive customers."

Alex scrambled through the presentation. There it was, slide twelve: a detailed strategy for automotive market penetration, complete with industry-specific recommendations that made no sense for Morrison's business.

"I think there might be some confusion in the analysis," Alex managed. "Let me dig deeper into these recommendations and get back to you with more specific insights."

The meeting ended awkwardly. Morrison was polite but clearly unimpressed.

Each AI tool had done exactly what Alex asked. The problem was that Alex had started treating AI responses like finished products instead of first drafts.

Back at the office, Alex studied the report the AI had produced, wondering how something so detailed could still feel so wrong.

The market analysis was technically accurate but based on data from the wrong industry. The competitor research was thorough but included companies that weren't actually competitors. The strategic recommendations were logical but completely irrelevant to Morrison's actual business.

Alex thought about calling Sarah to give her a heads-up but couldn't face that conversation yet.

The worst part wasn't embarrassing himself in front of Morrison. The worst part was the look Sarah would give him, the same look Alex's former boss had given him before the layoffs three years ago, right before everything fell apart. That look that said, 'Maybe we made a mistake.' Alex had promised Sarah this position would be different. Had promised the kids that this move to this city was worth uprooting them. Had promised himself that the career restart wasn't just desperation masquerading as reinvention.

Instead, Alex called Trent—a freelance consultant who'd been in the industry ten years longer and had become something of a mentor after they'd met at a franchise conference last year.

Alex called Trent that evening, feeling embarrassed.

"Let me guess," Trent said after hearing the story. "You got comfortable with AI results and stopped double-checking everything?"

"How did you know?"

"Because I've done it. We all do it. You get a few good results from ChatGPT or Claude, and suddenly you think you're an AI whisperer. Then AI confidently gives you completely wrong information and you look like an idiot."

"So what happened? I thought I understood how to communicate with these tools."

"You do understand communication. But you forgot that AI doesn't actually understand your business. It's incredibly good at sounding confident about things it knows nothing about."

Trent had a point. Alex had been so focused on improving prompts that the human oversight step had gotten lazy. The **CECH** approach worked, but only when you actually completed all four steps. Alex had mastered Context, Expectations, and Conversation but had started skipping Human oversight.

"Here's the thing about AI," Trent continued. "It's like having a research assistant who's read everything ever written but has never actually worked in your industry. They can find patterns, make connections, and generate ideas that sound brilliant. But they don't know which ideas are realistic for your specific situation."

That night, Alex reviewed every AI interaction from the past month. The pattern was clear: early successes had led to overconfidence, which had led to less verification, which had led to more mistakes that Alex had been too busy to catch.

The email that promised a delivery date Alex couldn't meet. The analysis that missed a key competitor because the AI didn't understand the industry landscape. The creative brief that sounded impressive but ignored the client's brand guidelines.

None of these were AI failures. They were human failures. Alex had stopped being the editor and started being just the user.

The next morning, Alex called Morrison.

"I owe you an apology," Alex said. "Yesterday's presentation included analysis that wasn't properly verified for your specific situation. I'd like to schedule a follow-up meeting with properly vetted recommendations."

"What happened?" Morrison asked. "It seemed like you really knew your stuff."

Alex took a breath. "I got overconfident in AI tools and didn't apply enough human judgment to the output. I know that sounds like making excuses, but I want to be transparent about what went wrong so we can fix it."

Morrison was quiet for a moment. "You know, I appreciate the honesty. Most consultants would have just sent me revised materials and pretended the first version never happened. Let's schedule that follow-up."

The second presentation took three times longer to prepare. Alex used the same AI tools but treated every output as a starting point, not a final answer. Every fact was verified. Every recommendation was tested against Morrison's actual business model. Every insight was filtered through Alex's understanding of Morrison's industry and goals.

The result was half the length but twice as valuable.

"Now this," Morrison said after the revised presentation, "feels like it was built specifically for our business. What changed?" "I remembered that AI is a tool for thinking, not a replacement for thinking," Alex replied.

The Alchemist's Insight

Early AI wins create dangerous overconfidence. When AI tools start producing good results, it's tempting to trust them completely. But AI confidence is not the same as AI accuracy. The better your prompts get, the more confident AI becomes about everything, including things it gets completely wrong. Success with AI isn't about eliminating human judgment. It's about knowing exactly when and how to apply it.

Your Next Step

For Individual Leaders: Audit your recent AI interactions. Look for places where you've started treating AI output as final instead of reviewing it critically. The more comfortable you get with AI, the more important human oversight becomes.

For Organizations: This is the moment to formalize your AI guardrails before overconfidence becomes expensive. Consider implementing:

- **AI Acceptable Use Policies (AUP)** that define when human verification is required
- **Tiered Review Standards** based on decision stakes (routine vs. high-impact)
- **Team Training on the CECH Framework** so everyone understands when to trust AI and when to verify
- **Regular AI Audits** where teams share "near-miss" stories like Alex's Morrison disaster

Chapter 4: The Human Element

Why People Still Hold All the Keys

The team meeting was going well until Alex mentioned the Morrison project.

"I think we should start using AI for all our client presentations," said Marcus, the newest team member. "Alex showed us how these tools can generate comprehensive market analysis in minutes. We could be way more efficient."

Alex shifted uncomfortably. Marcus hadn't heard about the Morrison disaster.

"Actually," Alex said, "I think we need to be more careful about how we use AI with client work."

"Careful?" Marcus looked puzzled. "But you said AI was a game-changer. You've been using ChatGPT for emails, Copilot for analysis, all that stuff."

Alex glanced around the conference room. The team was listening intently. This was exactly the conversation Alex had been dreading.

"AI is powerful," Alex began carefully. "But I learned something important last week. These tools can generate impressive content, but they don't understand our clients the way we do."

"What do you mean?" asked Jessica, the same Jessica who'd mastered AI tools across platforms. "My AI work has been pretty accurate."

Alex decided to share the Morrison story. The bankruptcy competitor. The automotive recommendations for a food service company. The confident-sounding analysis that was completely wrong.

"The problem wasn't the AI," Alex explained. "The problem was that I stopped thinking critically about what it produced. AI doesn't know that Morrison's biggest competitor went out of business. It doesn't understand their industry culture or their specific challenges. It just processes patterns in data."

Marcus frowned. "So we shouldn't use AI for client work?"

"That's not what I'm saying. I'm saying AI can't replace the human insight we bring to every project. It can help us think, research, and write. But it can't know what we know about our clients."

Sarah, the CEO, leaned forward. "Give me an example."

Alex pulled up the Morrison analysis on the screen. "Look at this recommendation about expanding into automotive markets. Technically, it's a well-structured growth strategy. The financial projections are logical. The market analysis is thorough. AI did exactly what I asked it to do."

"But?" Sarah prompted.

"But anyone who's worked with Morrison for five minutes knows they're passionate about food service. Their whole identity is built around feeding people. They'd never consider automotive because it's not who they are. AI doesn't understand passion or identity or company culture."

Jessica nodded slowly. "I see what you mean. When I use Copilot to analyze our accounting data, it finds patterns I might miss. But it doesn't know that the spike in March expenses was because we moved offices, or that the Q2 revenue dip was from delaying one client project. I have to provide that context."

"Exactly," Alex said. "AI sees the data. We see the story behind the data."

Marcus still looked skeptical. "But isn't that just a matter of better prompts? If we give AI more context about the client..."

Alex had wondered the same thing. "I tried that. I fed ChatGPT Morrison's company history, their mission statement, everything I could think of. It generated better recommendations, but they were still generic. AI can process information about Morrison, but it can't feel what it's like to walk into their offices and see how proud they are of their service awards."

The room was quiet for a moment.

"So what's the solution?" Sarah asked.

Alex had been thinking about this all week. "We use AI for what it's great at: research, analysis, first drafts, brainstorming. But we never forget that our job is to translate AI capabilities into human understanding. We're not AI users. We're translators between AI and our clients."

"That actually makes sense," Jessica said. "When I use Perplexity to research industry trends, it finds connections I wouldn't have thought to look for. But I still have to decide which trends actually matter to our specific clients and why."

Sarah smiled. "I like this framing. AI expands our capabilities, but human judgment is what makes those capabilities valuable."

Alex felt a wave of relief. The team was getting it.

"There's something else," Alex continued. "Our clients don't just want good analysis. They want to know that someone who understands their business created that analysis. They want the human touch."

Marcus raised his hand. "But what if AI gets better at understanding context and company culture?"

"Then we'll get better at being human," Alex replied. "The more AI can do, the more valuable we become at doing the things AI can't do. Building relationships. Understanding emotions. Making judgments based on experience, not just data."

After the meeting, Jessica approached Alex privately.

"Can I tell you something?" she said. "I've been using AI tools for months, and they've made me more productive. But the clients who keep coming back aren't impressed by my AI-generated reports. They keep coming back because they trust my judgment about what the reports actually mean."

Alex nodded. "That's it exactly. AI can help us be faster and smarter, but it can't help us be more trusted. That's still all human."

That night, Alex realized something important that shifted everything. For the first time since the Morrison disaster, hell, for the first time since accepting this job, Alex felt like an expert again rather than a fraud. Not because AI had gotten simpler, but because Alex finally understood the actual game. The Morrison failure wasn't just about overconfidence or skipping human oversight. It was about forgetting that clients hire people, not tools. And that insight, that hard-won, embarrassing, almost career-ending insight, was something no AI could have taught.

Morrison didn't want impressive AI-generated analysis. Morrison wanted Alex's expertise, enhanced by AI but filtered through years of business experience and human understanding.

The real competitive advantage wasn't using AI better than competitors. It was using AI while remaining more human than competitors.

The Alchemist's Insight

AI amplifies human capabilities but can't replace human understanding. The more powerful AI becomes, the more valuable human insight, judgment, and relationships become. Your job isn't to become an AI expert, it's to become a better human who happens to use AI.

Clients don't buy AI-generated solutions. They buy human judgment that's been enhanced by AI capabilities.

Your Next Step

Identify what makes your judgment unique in your role or industry. Those insights, relationships, and experiences are your competitive advantage. Use AI to amplify these human strengths, not replace them. The goal is becoming irreplaceably human, not artificially intelligent.

AI handles the data. You handle the meaning behind the data.

Chapter 5: The AI Sherpa

An Unlikely Guide to Mastery

Alex was running late for the quarterly franchise conference. Again.

The company had grown to sixty franchise locations over the past five years, and these quarterly meetings with franchisees were always a balancing act. Each location had different challenges, different markets, different personalities running them. As Director of Franchise Operations, Alex's job was to help corporate initiatives land successfully across this diverse network, but that meant understanding what each franchisee actually needed, not just what corporate wanted to roll out.

Keeping everyone aligned while respecting their independence was like conducting an orchestra where every musician wanted to play a different song.

Today's agenda was supposed to focus on the new AI tools corporate was rolling out. Customer service chatbots, automated scheduling systems, inventory management AI. After the Morrison project disaster a few weeks ago, Alex was particularly nervous about how to position AI as helpful rather than risky.

The franchisees were... skeptical.

"I don't need a robot talking to my customers," Tom from the Denver location had said in the pre-meeting survey. "People come here because they want human service."

"What happens when the AI breaks down?" asked Maria from Phoenix. She managed one of their highest-volume locations but was notoriously technology-averse. "My staff can barely handle the current system updates."

Alex understood their concerns completely now, especially after watching AI confidently provide wrong information about competitors and automotive recommendations for food service companies during the Morrison presentation. These were seasoned business owners who'd built their success on personal relationships, local market knowledge and following their franchise systems. How could Alex ask them to trust AI when Alex had just learned not to trust it blindly?

The conference room was buzzing with the usual pre-meeting energy when Alex noticed someone new sitting in the back corner. A woman in her fifties, laptop open, quietly observing the room with the kind of attention that suggested she was learning more from watching than most people learned from talking.

"That's Maya Chen," whispered David, the corporate operations manager. "Sarah brought her in as the company's new data strategist and consultant. Supposed to help us figure out this AI thing for franchise operations."

Another consultant. After Dr. Wellington and the parade of AI experts, Alex suppressed a groan.

But as the meeting progressed, Maya didn't behave like the other consultants Alex had encountered. She didn't present slides about neural networks, gimmics, or transformation paradigms. Instead, she asked questions. She had a calm confidence that contrasted sharply with the tech buzz around her, and she listened more than she talked.

"Tom," she said during the discussion about customer service AI, "tell me about your best customer interaction last week."

Tom looked surprised but launched into a story about helping a regular customer find the right product for her specific needs. "She'd tried three other places, but nobody took the time to really understand what she was looking for."

Maya nodded. "And what did you know that the other places didn't?"

"Well, I knew she was dealing with a specific problem. I knew her budget constraints. And I knew she'd rather pay a little more for something that actually works than keep buying cheap solutions that fail."

"Right," Maya said. "So when we talk about AI for customer service, we're not talking about replacing that kind of interaction. We're talking about handling the routine stuff so you have more time for the interactions that really matter."

The room shifted. Tom leaned forward.

Alex recognized what was happening. Maya was doing exactly what Alex had learned to do with AI tools, providing context and clear expectations. But she was doing it in reverse: understanding the human context first, then figuring out where AI could fit.

"What do you mean?" Tom asked.

Maya pulled up her laptop and projected it on the screen. "What if AI handled appointment scheduling, basic product questions, and inventory checks? Not the relationship building or complex problem solving, but the repetitive tasks that currently take up 60% of your customer service time?"

"That... might actually help," Tom admitted.

Maya turned to Maria. "You mentioned concerns about system reliability. What's your biggest frustration with technology updates?"

"They never work the way they're supposed to," Maria said. "And when something goes wrong, we can't get help that makes sense. The support team assumes we know things we don't know."

Alex winced, remembering how overconfident AI responses had led to the Morrison presentation disaster.

"So if we implement AI tools," Maya continued, "the most important thing isn't the AI itself. It's making sure you have clear, simple ways to fix problems when they happen. Human backup systems that actually work."

Alex watched, fascinated. Maya wasn't selling AI solutions. She was translating AI capabilities into franchisee language. She understood that success wasn't about the technology, it was about helping business owners stay focused on what made their locations successful while removing the obstacles that prevented them from doing their best work.

During the break, Alex approached Maya.

"I have to ask, how do you do that? I've been struggling with AI for months. I learned how to communicate better with ChatGPT and Copilot, even got decent at using multiple AI platforms. But I also just had a major failure where I trusted AI too much and embarrassed myself in front of a client. Yet you just made AI make sense to a room full of sixty skeptical franchise owners in twenty minutes."

Maya smiled. "What do you think I did?"

Alex thought for a moment. "You didn't talk about AI. You talked about their businesses. You showed them how AI could help them be better at what they're already good at."

"Exactly. And what else?"

"You... asked about their problems first, instead of starting with solutions. Kind of like setting context and expectations, but for humans instead of AI."

"Right. And?"

Alex was starting to see the pattern. "You treated AI like a tool to solve their specific challenges, not as something they needed to learn for its own sake. And you kept humans in control of the important stuff."

"One more thing," Maya prompted.

Alex looked back at the room, where Tom was now enthusiastically discussing AI applications with Maria. "You helped them realize they were already doing the hard part, building relationships and understanding customers. AI just handles the easy stuff they don't want to spend time on anyway."

Maya nodded. "That's the foundation of everything. AI isn't about replacing human expertise. It's about freeing up human expertise to focus on what humans do best."

"But how do you know so much about franchise operations? And how did you learn to think about AI this way?"

Maya laughed. "I'm not a franchise expert. I learned this approach from Ford Saeks, a business strategist who's been helping leaders navigate technological transformation for decades. He works with all kinds of organizations—franchises, corporations, associations—helping them unleash AI innovations while avoiding the pitfalls and keeping the human experience central to everything they do."

Alex perked up. "Ford Saeks? The keynote speaker? I think I've seen his name on some conference programs."

"That's him. He wrote *AI Mindshift* and speaks at events worldwide. His whole philosophy is that successful AI adoption isn't about becoming more technical, it's about becoming more strategic and more human. He taught me to be a translator between AI capabilities and human needs."

"A translator?"

"Think about it this way," Maya said. "Each of your franchisees is already an expert in their market, their customers, their operations. Ford's approach isn't about making them AI experts. It's about showing them how AI can amplify what they already know how to do."

Alex felt something click into place. "So success with AI isn't about mastering the technology. It's about understanding how to integrate it into expertise that already exists."

"Now you're getting it," Maya said. "Ford calls people like me 'AI Sherpas,' guides who help others navigate the terrain without getting lost in the technical complexity. Want to learn how to think about AI this way? Not just for your franchise network, but for any business challenge?"

For the first time in months, Alex felt hopeful about AI. Not the desperate hope that had driven those 2 AM YouTube tutorial binges, but something steadier. Something that felt less like panic and more like possibility. Maybe this job wasn't going to end like the last one. Maybe the risk of uprooting the kids wasn't a mistake. Maybe—and this was the thought that caught

Alex by surprise—maybe this was exactly where Alex was supposed to be all along.

"I'd like that very much," Alex said.

The Alchemist's Insight

AI can analyze patterns, but it takes human wisdom to interpret them. The Human Element is not optional; it is the catalyst that turns data into decisions

The best AI guides aren't technology experts, they're translation experts. They understand how to bridge the gap between what AI can do and what you already do well. Success comes from amplifying existing expertise, not replacing it with artificial intelligence.

AI mastery isn't about learning new skills. It's about applying AI to the skills you've already developed.

Your Next Step

Look for AI guidance that starts with your current expertise, not with AI capabilities. The right approach helps you become better at what you're already good at, rather than forcing you to become good at something entirely new.

Find translators who speak your language, not AI's language.

For leadership teams: The frameworks Maya shared with Alex, the CRAFT method, Speed-Check System, and human-AI collaboration principles, aren't theoretical. They're battle-tested approaches that work across industries because they start with human capability and add AI strategically.

Bring this transformation to your organization. Ford Saeks delivers keynote presentations and interactive training that help teams move from AI confusion to confident mastery. Watch his speaker video and explore his AI training programs at **ProfitRichResults.com**, where you'll find resources designed for leaders who want to harness AI without losing their humanity.

PART II:
THE
FRAMEWORK FOR
TRANSFORMATION

Chapter 6: The Alchemist's First Rule

Why Purpose Beats Technology Every Single Time

Maya had suggested they meet at a coffee shop instead of the corporate conference room. "Less formal," she'd said. "Better for real conversation."

Alex arrived early, still processing everything from the franchise meeting. Watching Maya work with sixty skeptical business owners had been like watching someone speak a different language, one that everyone somehow understood perfectly.

Maya arrived with a worn notebook and that same quiet confidence Alex had noticed before.

"So," Maya said, settling into her chair, "tell me about your worst AI failure."

Alex winced. "Morrison. I used AI to create what I thought was impressive market analysis and strategic recommendations. Turned out the AI had recommended automotive expansion strategies for a food service company and cited a competitor that had filed for bankruptcy."

"What went wrong?"

"I got overconfident. I'd learned how to communicate better with ChatGPT, Copilot, all these different platforms. My prompts were getting better results, so I started trusting the outputs without really thinking about them."

Maya nodded. "And before that? What was your first AI frustration?"

"The tools never gave me what I actually needed. I'd ask for a professional email and get corporate-speak. I'd ask for analysis and get generic insights that missed the point."

"Two different problems," Maya observed. "First, you couldn't get AI to understand what you wanted. Then, you got too comfortable with what AI was giving you. What's the connection?"

Alex thought for a moment. "Both times, I was focused on the wrong thing?"

"Keep going."

"The first time, I was focused on the tool. How to prompt it, how to get better outputs. The second time, I was focused on efficiency. How to save time, how to get more done."

Maya smiled. "And what should you have been focused on?"

Alex paused, thinking about the Morrison disaster. "The outcome I was actually trying to achieve?"

"Exactly. Let me ask you this: what was the real purpose of that Morrison presentation?"

"To show them we understood their business and could help them grow."

"And did AI help you achieve that purpose?"

"No. It made me look like I didn't understand their business at all."

Maya opened her notebook and wrote something down. "This is what Ford Saeks calls the Alchemist's First Rule. **Purpose beats technology every single time.**"

She turned the notebook toward Alex. At the top of the page, she'd written:

Start with WHY you're trying to accomplish something, not HOW you're going to accomplish it.

"Ford has been teaching this principle for decades," Maya continued. "Most people approach AI backwards. They start with the technology and try to figure out what to do with it. The alchemists of old did the opposite. They started with what they wanted to transform and then figured out which processes would get them there."

Alex looked at the note. "So instead of asking 'How can AI help me write better emails,' I should ask 'What am I trying to accomplish with this email?'"

"Right. And once you know the purpose, AI becomes one of many tools you might use to achieve it. Not the tool you have to use."

"Can you give me an example?"

Maya pulled out her phone. "Yesterday, a client asked me to help her use AI for social media content. So I asked her: what's the purpose of your social media?"

"What did she say?"

"First she said 'to get more followers.' So I asked, 'Why do you want more followers?' Then she said 'to increase brand awareness.' So I asked, 'Why do you want to increase brand awareness?'"

Alex could see where this was going. "What was the real purpose?"

"To build trust with potential customers so they'd be more likely to buy from her instead of competitors."

"That's completely different from getting followers."

"Exactly. And once we understood the real purpose, the AI strategy became obvious. Instead of using ChatGPT to generate generic inspirational posts, we used it to help her share specific customer success stories and industry insights. Same AI tool, completely different approach, much better results."

Alex felt something click. "So the Morrison presentation problem wasn't that I used AI. It was that I lost sight of the real purpose."

"What was the real purpose?"

"Building trust and demonstrating understanding. But I got distracted by creating impressive-looking analysis."

Maya nodded. "AI is incredibly good at creating impressive-looking things. It's terrible at knowing whether those things serve your actual purpose."

"So how do I avoid that trap?"

Maya flipped to a new page in her notebook. "Ford taught me to ask three questions before opening any AI tool: "

She wrote:

1. *What am I trying to achieve?*

2. *Why does this matter to the person I'm serving?*

3. *How will I know if I've succeeded?*

"Once you have those answers, AI becomes much more useful. It helps you achieve your purpose instead of distracting you from it."

Alex studied the questions. "This feels like it would work for more than just AI."

"That's Ford's point. Purpose-driven thinking works for everything. AI just makes it more important because AI can produce so much content so quickly that you can waste a lot of time going in the wrong direction."

They sat quietly for a moment, Alex absorbing the lesson.

"Ford has a saying," Maya added. "'The most powerful technologies amplify human purpose... they don't replace it.' He's been helping leaders navigate technological change for thirty years, and he always comes back to this: successful AI adoption isn't about becoming more artificial, it's about becoming more intentionally human."

"More intentionally human?"

"Being clearer about what you're trying to accomplish and why it matters. AI can help you execute faster and more efficiently, but only humans can determine what's worth executing in the first place."

Alex felt a shift in thinking. "So the Alchemist's First Rule is really about staying human while using artificial intelligence."

"Now you're getting it," Maya said. "Ready to learn how Ford applies this to everything he does with AI?"

The Alchemist's Insight

Purpose beats technology every single time. Before using any AI tool, get clear on what you're trying to achieve and why it matters. AI amplifies human purpose, it doesn't replace it. The most powerful AI applications start with human intention, not technological capability. .

Your Next Step

Before your next AI interaction, answer three questions: What am I trying to achieve? Why does this matter to the person I'm serving? How will I know if I've succeeded? Let purpose drive your AI usage, not the other way around.

Technology serves purpose, not the reverse.

Chapter 7: Crafting Gold

The Secret to Making AI Actually Deliver Results

A week later, Alex was back at the coffee shop, this time with a notebook and a specific challenge.

"I tried your three questions on a project yesterday," Alex told Maya. "It helped me focus on the real purpose, but I'm still struggling with getting AI to actually deliver what I need."

Maya smiled. "Let me guess. You know what you want to achieve, but the AI outputs still feel generic or miss the mark?"

"Exactly. I asked Copilot to help me analyze our franchise performance data. I was clear about my purpose, which was to identify which locations needed support and which were ready for expansion. But the analysis came back with standard metrics that didn't tell me anything I didn't already know."

"Show me exactly what you asked for."

Alex pulled out his phone and read: "Analyze this franchise performance data and identify locations that need support versus those ready for expansion."

Maya nodded. "Good start. You were clear about purpose. Now we need to talk about crafting gold. That's Ford's approach to getting AI to produce results that are not just useful, but truly valuable."

"Crafting gold?"

"The alchemists believed they could transform base metals into gold through the right process. Ford teaches that you can transform generic AI outputs into valuable insights through the right communication approach. It's not just about better prompts, it's about better conversations."

Maya opened her notebook to a page filled with examples. "Ford has developed what he calls the **CRAFT method.** He actually recommends using AI to help write more effective prompts using this framework."

It stands for Context, Role, Audience, Format, and Tone.

CONTEXT ROLE AUDIENCE FORMAT TONE

C - Context (What's the situation or objective?)

R - Role (Who are you asking AI to be?)

A - Audience (Who is this for or use the result?)

F - Format (What structure or output do you need?)

T - Tone (How should it sound or feel?)

"Here's what Ford suggests," Maya continued. "First, use AI to help you build better prompts with CRAFT, then use those improved prompts to get better results. It's like using AI to teach itself how to help you better."

"How does that work?"

Maya pulled out her phone and activated the voice assistant. "Let me show you. I'll ask ChatGPT using voice to help me create a better prompt for your franchise analysis using the CRAFT method."

She spoke clearly into her phone: "Help me create an effective prompt using the CRAFT method. That's Context, Role, Audience, Format, and Tone. I need to analyze franchise performance data to identify which of sixty locations need support versus which are ready for expansion. The analysis is for executives who need actionable insights to make strategic decisions."

The voice response came back with a structured breakdown that transformed Alex's original simple request into a comprehensive prompt framework.

"See how AI helped structure a much better request?" Maya said. "Ford discovered that AI is actually great at helping you communicate more effectively with AI, whether you're typing or talking to it. Voice prompting is especially useful because you can think out loud and refine your request naturally."

Maya continued the voice conversation with her AI: "Now take that CRAFT framework you just created and write the actual prompt I should use for the franchise analysis."

The AI responded with a detailed, structured prompt that was far more comprehensive than Alex's original request.

"This is clever," Alex said. "So I use AI to help me ask AI better questions, and I can do it by voice when I'm thinking through the problem?"

"Exactly. Ford calls it '**meta-prompting.**' You're not just using AI tools, you're using AI to get better at using AI tools. Voice makes it even more natural because you can have a real conversation about what you're trying to accomplish."

Alex tried the approach, first speaking to ChatGPT to help him craft a better prompt using CRAFT, then using that improved prompt for the actual analysis. The difference was dramatic. Instead of generic performance metrics, the AI produced insights about market-specific challenges, seasonal patterns affecting certain locations, and specific recommendations tied to observable data trends.

"This is exactly what I needed," Alex said, scrolling through the results. "And using voice to work through the CRAFT method felt more natural than typing it all out."

"That's the power of crafting gold," Maya said. "But there's one more piece Ford emphasizes; the conversation doesn't end with the first response."

"What do you mean?"

"AI gives you a starting point, not a final answer. The real value comes from the back-and-forth. Ford calls it 'collaborative intelligence,' and you provide human judgment, AI provides capability, and together you create something neither could produce alone."

Maya pointed to Alex's analysis. "What questions does this raise for you?"

Alex scanned the results. "Well, AI identified three locations as 'ready for expansion' but I know one of them just lost their lease. And it flagged two locations as needing support, but I think the issues might be more about local market conditions than operational problems."

"Perfect. So what's your next move?"

"Keep the conversation going. Ask AI to revise the analysis with this additional context."

"Exactly. Ford says the magic happens in the iteration. The first response gives you material to work with. The second response, informed by your expertise, gives you insights you can actually use."

Alex used voice again to add context about the lease situation and market conditions, then asked AI to revise its recommendations. The conversational approach felt natural, and the updated analysis was significantly more accurate and actionable.

"This feels like having a really smart research assistant who needs guidance but can process information faster than I ever could," Alex said. "And being able to talk through the refinements makes it feel more like actual collaboration."

Maya smiled. "Ford has a phrase for exactly that: 'Treat AI like your most capable intern who is brilliant at execution, terrible at judgment.' The CRAFT method helps you provide the judgment while AI handles the execution, whether you're typing or talking."

"So crafting gold isn't about finding the perfect prompt?"

"It's about having better conversations that combine AI's processing power with human understanding. Ford's spent decades helping people navigate technological change, and he always comes back to this: the technology amplifies the quality of your thinking, but it can't replace the thinking itself."

Alex looked at the revised analysis, which now included nuanced recommendations that accounted for local market realities and operational constraints. "I could never have processed this much data this quickly on my own, but AI could never have made these judgments without understanding our specific situation."

"Now you're thinking like an AI alchemist," Maya said. "Ready to learn how Ford balances AI speed with human wisdom?"

The Alchemist's Insight

Great AI results come from great conversations, not great prompts.

Clear input creates clear output. When you use the **CRAFT Method** to set context, role, audience, format, and tone, you turn vague ideas into precise results.

Use AI to help you craft better prompts, whether by voice or text, then engage in iterative conversation where human judgment guides AI capability. Voice prompting makes the collaboration feel more natural and allows you to think through problems conversationally.

Your Next Step

Try using AI to improve your AI interactions: Use voice or text to ask ChatGPT or your preferred tool to help you create better prompts using the **CRAFT Method**. Then use those improved prompts for better results, and treat the first response as a starting point for deeper conversation.

The best AI results come from collaborative intelligence, not artificial intelligence alone.

Chapter 8: Balancing the Scales

Human Judgment Meets AI Speed

"I have a confession," Alex said as they settled into their usual corner at the coffee shop. "I've been using your **CRAFT method** for everything this week, and the results have been incredible. But yesterday, I almost made a huge mistake."

Maya raised an eyebrow. "What happened?"

"I was rushing to prepare for a board presentation about our franchise expansion strategy. I used AI to analyze market data, create financial projections, even draft the presentation slides. Everything looked perfect. Professional, comprehensive, data-driven."

"Sounds successful."

"That's what I thought. But then I realized I hadn't actually verified any of the market data. AI had pulled information that looked authoritative, but when I double-checked, some of the key statistics were from 2019, and one of the 'growth markets' it recommended had actually seen a 30% population decline in the past two years."

Maya nodded knowingly. "You discovered the speed trap."

"The speed trap?"

"Ford talks about this constantly in his work with organizations. AI's greatest strength is speed and that can also become its greatest weakness when it outpaces human judgment. The faster AI gets, the more important it becomes to slow down and think."

Alex leaned forward. "But isn't the whole point to be more efficient?"

"Efficient at the right things, yes. But Ford makes a crucial distinction: there's a difference between moving fast and moving smart. AI helps you move fast. Human judgment helps you move smart."

Maya opened her notebook to a diagram she'd drawn. "Ford uses the metaphor of scales, like the ones ancient alchemists used to measure ingredients. On one side, you have AI speed and capability. On the other, you have human wisdom and judgment. The goal isn't to choose one or the other. It's to keep them in balance."

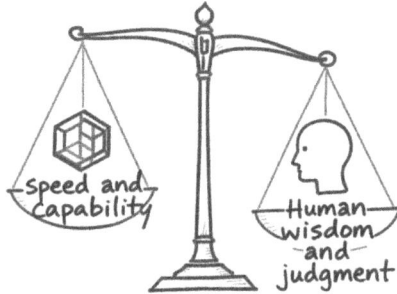

She pointed to the diagram. "When AI speed outweighs human judgment, you get impressive-looking results that are fundamentally flawed just like your market analysis with outdated data."

"And when human judgment outweighs AI capability? You get perfect decisions that come too late to matter. Ford has seen companies spend months analyzing problems that AI could have helped them understand and solve in hours."

Alex studied the drawing. "So how do you maintain the balance?"

"Ford has developed what he calls the **'Speed-Check System.'** It helps you decide when to trust AI output and when to pause for human review"

Step 1 – Assess importance: Does this decision affect money, reputation, or people?

Step 2 – Verify accuracy: Has the information been cross-checked with reliable sources?

Step 3 – Confirm context: Does the output align with your purpose and brand?

Step 4 – Add human review: Someone must validate before acting.

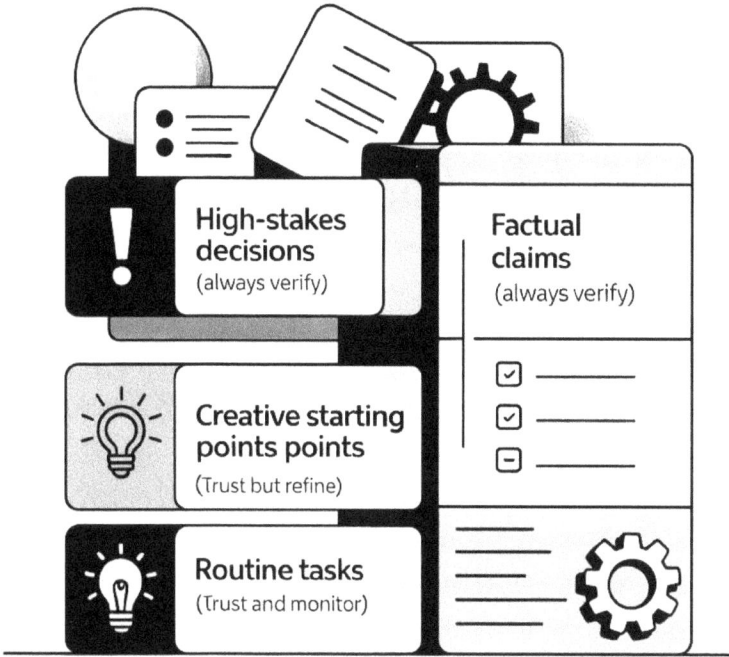

High-stakes decisions (always verify)	Factual claims (always verify)
Creative starting points points (Trust but refine)	☑ —— ☑ —— ⊟ ——
Routine tasks (Trust and monitor)	

"The key insight Ford discovered," Maya continued, "is that the stakes determine the balance. For routine tasks like scheduling or basic data organization, you can trust AI and monitor the results. For high-stakes decisions or factual claims, you always verify."

"That makes sense. My board presentation was definitely high-stakes, and I was making factual claims about market data."

"Exactly. Ford says the mistake most people make is treating all AI outputs the same way. They either trust everything or verify everything. Both approaches miss the point."

Maya pulled out her phone and demonstrated. "Watch this. I'm going to ask AI to help me plan a team lunch for next week."

She spoke to her phone: "Help me plan a team lunch for eight people next Tuesday. Suggest three restaurant options in downtown, include approximate costs and whether they accommodate dietary restrictions."

The AI quickly provided three restaurant suggestions with details about pricing and dietary options.

"Now," Maya said, "this is a routine task with low stakes. I can trust this information and just monitor the results. If the restaurant doesn't work out, it's not a crisis."

She then asked: "What's the current inflation rate in the United States and how is it affecting restaurant pricing?"

The AI provided statistics and analysis about inflation and restaurant costs.

"This," Maya said, "involves factual claims that could affect business decisions. Before I use this information for anything important, I'd need to verify the sources and check if the data is current."

Alex was starting to see the pattern. "So the system isn't about trusting AI or not trusting AI. It's about matching the level of verification to the level of importance."

"Ford puts it this way: 'Trust AI with your time, verify AI with your reputation.'"

Alex laughed. "That's good. So anything that could affect my reputation or my company's reputation gets verified, but routine efficiency tasks get trusted."

"Right. And Ford emphasizes that this balance isn't just about individual tasks. It's about building organizational systems that maintain the right speed-judgment balance."

"What does that look like?"

Maya turned to another page in her notebook. "Ford works with companies to establish what he calls 'AI guardrails'—systematic checkpoints that ensure human judgment stays in the loop for decisions that matter."

"The genius of this system," Maya said, "is that it lets you move at AI speed for most things while slowing down for the things that really matter. You're not bottlenecking everything through human verification, but you're not blindly trusting AI either."

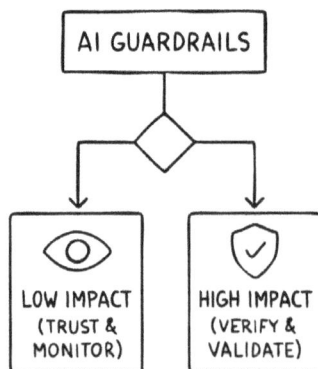

AI GUARDRAILS

LOW IMPACT (TRUST & MONITOR)

HIGH IMPACT (VERIFY & VALIDATE)

Alex thought about his near-miss with the board presentation. "I wish I'd had this framework yesterday. I would have caught the problem before almost embarrassing myself in front of the board."

"Ford always says that AI's biggest gift isn't just making us faster, it's forcing us to think more clearly about what deserves our careful attention and what doesn't."

Maya closed her notebook. "The ancient alchemists spent years learning to balance their ingredients perfectly. They knew that too much of any element, even the right elements, could ruin the entire process. AI alchemy works the same way."

"So becoming an AI alchemist isn't just about using AI better. It's about becoming more intentional about when and how I apply human judgment."

"Now you're getting it," Maya said. "Ready to learn how Ford helps organizations avoid the ethical pitfalls that come with all this power?"

The Alchemist's Insight

AI's greatest strength can become its greatest weakness when speed outpaces judgment. The goal isn't choosing between AI efficiency and human wisdom, it's maintaining the right balance. Trust AI with your time, verify AI with your reputation.

Use the **Speed-Check System:** High-stakes decisions and factual claims always get verified. Creative starting points get trusted but refined. Routine tasks get trusted and monitored.

Your Next Step

Assess your recent AI interactions: Which ones were high-stakes or involved factual claims that should have been verified? Which ones were routine tasks that could be trusted? Develop your own **Speed-Check System** based on what could affect your reputation versus what just affects your efficiency.

Balance speed with wisdom, not speed versus wisdom.

Chapter 9: Avoiding the Dark Arts

Building Trust When Everyone Thinks AI Will Replace Them

"We need to talk," said Marcus, the team's newest member, cornering Alex after the Monday morning leadership meeting. "People are freaking out about this AI stuff."

Alex had been dreading this conversation. Ever since the company started implementing Maya's AI frameworks, productivity had increased dramatically. Customer response times were faster. Data analysis was more thorough. Even routine tasks were getting done with impressive efficiency.

But the team was scared.

"Jennifer from customer service thinks AI is going to take her job," Marcus continued. "She heard about companies replacing entire support teams with chatbots. And David in sales is convinced that AI agents will start handling client relationships directly."

Alex sighed. This was the dark side of AI success. The better the tools worked, the more people worried about becoming unnecessary.

"What did you tell them?"

"I didn't know what to say. I mean, AI is getting pretty sophisticated. Voice calls, image analysis, web browsing, workflow automation. How do we know what's actually real and if it will eventually replace us?"

That afternoon, Alex called Maya.

"We have a trust problem," Alex explained. "The team is afraid that learning to use AI better is basically training their own replacements."

Maya was quiet for a moment. "Ford talks about this constantly in his keynotes. He has a saying: 'AI won't replace humans... it will replace humans not using AI.'"

"What does that mean?"

"Think about it this way. The real threat isn't AI taking your job. The threat is someone else in your industry using AI to do your job better, faster, and more efficiently than you can without it."

Alex felt a chill. "So we're in an arms race?"

"Not an arms race, it's an adaptation race. Ford always says the greatest fear about AI isn't adopting it too soon, it's waiting too long and falling behind. The companies and individuals who figure out how to work with AI will have a massive advantage over those who resist it."

"But what about all the predictions about AI replacing millions of jobs?"

"Ford's honest about this," Maya said. "AI will definitely replace millions of jobs. But it will also create new ones. The question isn't whether change is coming, it's whether you'll be prepared for it."

Maya paused. "Ford has an interesting perspective on this as a keynote speaker. AI can generate presentations, analyze audiences, even create talking points. But he's not worried about being replaced."

"Why not?"

"Because he believes there will be a shift toward events and interactions without technology, where human-to-human connection becomes a premium experience. The more automated our world becomes, the more valuable authentic human interaction becomes."

That made sense to Alex. "So how do I help my team see AI as an opportunity instead of a threat?"

"Ford has a framework for this. He calls it "Avoiding the Dark Arts." He wasn't talking about magic. He meant the seductive shortcuts and hype that promise results without responsibility The dark arts are the mistakes leaders make that create fear instead of confidence.

"What kind of mistakes?"

"The biggest one is pretending AI won't change anything, or promising that nobody's job will be affected. Ford says transparency builds trust better than false reassurance."

Maya opened her notebook to a page titled "**Trust-Building Principles.**"

"Ford's approach is to be honest about change while helping people see their irreplaceable value. Here's what he recommends:"

She read from her notes:

| Acknowledge the real changes | Focus on human amplification | Involve people in the process | Highlight human–only value |

1. Acknowledge the Real Changes: Don't pretend AI won't affect jobs, because it will. But help people understand how their roles will evolve, not disappear.

2. Focus on Human Amplification: Show how AI makes people better at their current jobs before talking about efficiency gains.

3. Involve People in the Process: Let the team help decide how AI gets implemented rather than imposing solutions from above.

4. Highlight Human-Only Value: Identify what each person brings that AI can't replicate.

"Can you give me an example?" Alex asked.

"Sure. Let's take Jennifer in customer service. Instead of saying 'AI won't replace you,' try Ford's approach: 'AI will handle the routine questions so you can spend more time solving complex problems and building relationships with frustrated customers. That is the work you are already great at but never have enough time for.'"

Alex could see the difference. "That's not about protecting her job from AI. That's about making her job better with AI."

"Exactly. Ford says the goal isn't to eliminate the fear of change, it's to redirect that energy toward adaptation and growth."

"What about the people who are genuinely at risk? Some jobs really will be replaced."

Maya nodded. "Ford's honest about this too. Some roles will disappear, but new ones will emerge. The key is helping people develop skills that complement AI rather than compete with it."

"Like what?"

"Emotional intelligence, creative problem-solving, relationship building, strategic thinking. Ford points out that as AI handles more routine cognitive tasks, human skills like empathy, intuition, and complex judgment become more valuable, not less."

Alex thought about the franchise meeting where Maya had worked with sixty skeptical business owners. "That's what you did with our franchisees. You didn't compete with AI's capabilities, you showed them how to direct those capabilities toward human goals."

"Right. Ford's philosophy is that the future belongs to people who can collaborate with AI, not people who can compete with it or hide from it."

The next morning, Alex called a team meeting.

"I want to address the elephant in the room," Alex began. "I know there are concerns about AI and job security. I'm not going to tell you that nothing will change, because that wouldn't be honest."

The room was tense.

"AI will change how we work. Some tasks you do today will be handled by AI tomorrow. But here's what I've learned: the goal isn't to make AI unnecessary. The goal is to make ourselves irreplaceable by using AI to amplify what we're already great at."

Alex looked around the room. "Jennifer, you're amazing at calming down frustrated customers and finding creative solutions to their problems. AI can handle the routine questions so you have more time for the situations where you really make a difference."

Jennifer looked skeptical but interested.

"David, your strength is understanding what clients really need, even when they can't articulate it clearly. AI can help you research their industries and prepare better, but it can't read between the lines the way you do."

Alex continued around the room, helping each person see how AI could enhance their unique strengths rather than replace them.

"Here's what Ford Saeks teaches," Alex said, remembering Maya's guidance. "AI won't replace humans. It will replace humans not using AI. The question isn't whether to adapt to this technology. The question is whether you want to be in control of how you adapt."

The room was quiet for a moment.

"So what's our next step?" Jennifer asked.

"We learn together. We experiment with AI tools that make your specific jobs better. And we remember that the most valuable thing about each of you isn't what you know, it's how you think, how you connect with people, and how you solve problems that don't have obvious answers."

Marcus raised his hand. "What if we're wrong? What if AI eventually can do all of that too?"

Alex smiled, remembering Maya's words. "Then we'll focus on becoming even more human. The more artificial intelligence can do, the more valuable authentic human intelligence becomes."

The Alchemist's Insight

Fear of AI replacement creates paralysis that guarantees replacement. The real threat isn't AI taking your job, it's someone else using AI to do your job better. Build trust by being honest about change while helping people discover their irreplaceable human value.

AI won't replace humans. It will replace humans not using AI.

Your Next Step

Have honest conversations about how AI will change work, not whether it will change work. Help your team identify their unique human strengths, then show them how AI can amplify those strengths. The goal is becoming irreplaceable through adaptation, not irreplaceable through resistance.

Embrace change to control change. Resist change to be controlled by change.

Chapter 10: Continuous Refinement

How to Stay Ahead While Your Competitors Chase Shiny Objects

Three months after implementing Maya's frameworks, Alex received an unexpected email from a competitor. The past quarter had been an experiment in Ford's Continuous Refinement principle, testing, learning, and improving their AI approach a little each week.

"Saw your company's customer response times have improved significantly," wrote David Kim, VP of Operations at Meridian Corp. "We're implementing GPT-25 next quarter and expecting similar results. What AI platform are you using?"

Alex stared at the email. GPT-25? Alex hadn't even heard of GPT-25.

A quick web search revealed that a new AI model had indeed been released, promising revolutionary improvements in reasoning, multimodal capabilities, and autonomous task execution. LinkedIn was buzzing with posts about "upgrading to GPT-25" and "the competitive advantage of next-generation AI."

Alex felt a familiar pang of anxiety. Were they already falling behind?

That afternoon, Alex called Maya.

"Should we be worried about GPT-25?" Alex asked. "Our competitors are already planning upgrades, and I'm wondering if our current AI approach is about to become obsolete."

Maya laughed. "Let me ask you something. How are your current AI implementations working?"

"Really well, actually. Customer service response times are down 40%. Our franchise performance analysis is more accurate and actionable than ever. The team is using AI confidently across multiple tasks."

"And what specific AI models are you using for these results?"

Alex paused. "Honestly? I'm not sure. We have people using ChatGPT, Copilot, Claude, voice assistants, some image analysis tools. But we're focused more on how we use them than which version we're using."

"That's exactly right. You're practicing what Ford calls **'Continuous Refinement'**—the difference between AI alchemists and AI chasers."

"What's the difference?"

"AI chasers upgrade their tools constantly but never upgrade their thinking. AI alchemists upgrade their thinking constantly and only upgrade their tools when it serves a specific purpose."

She paused, letting it sink in. "It's the same trap leaders fall into every tech cycle, confusing activity with progress."

Maya explained Ford's approach. "He's been watching this pattern for decades across every technology wave. There are always early adopters who chase the latest features, and there are always strategic adopters who focus on sustainable improvement."

"Which approach wins?"

"Ford has seen this play out repeatedly. The feature chasers get temporary advantages but burn out from constant change. The strategic refiners build sustainable competitive advantages that compound over time."

"So should we ignore GPT-25?"

"Not ignore... evaluate. Ford's Continuous Refinement framework helps you decide when new AI capabilities are worth adopting and when they're just distractions."

Maya opened her notebook to a page titled "**The Refinement Filter.**"

"Before adopting any new AI capability, Ford asks four questions:"

> **1. Real Problem First:** Does this solve a problem we actually have?

> **2. Enhancement or Distraction:** Will this make our existing AI work better or just create more AI work?

> **3. Integration Reality Check:** Can our team adopt this without disrupting what's already working?

> **4. Customer Impact Test:** Will this help us serve customers better or just make us look more cutting-edge?

"It's like a filter," Maya continued. "New AI capabilities that pass all four questions get considered. Everything else gets ignored, no matter how impressive the marketing sounds."

Alex thought about GPT-25. "So I should ask whether upgrading would actually solve problems we have, not just give us better AI?"

"Exactly. Ford says the most successful AI implementations aren't about having the latest technology, they're about continuously improving how you use whatever technology you have."

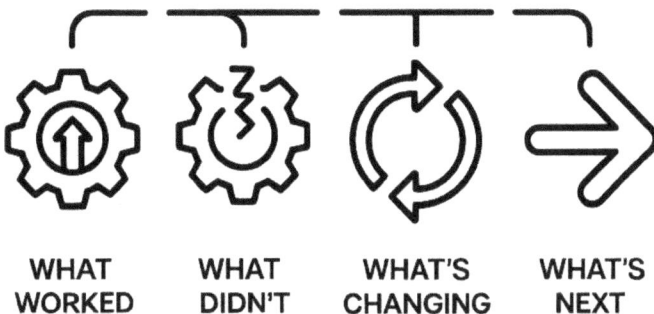

WHAT WORKED   WHAT DIDN'T   WHAT'S CHANGING   WHAT'S NEXT

"What does that look like practically?"

Maya flipped to another page. "Ford calls it the **'Weekly Refinement Ritual**.' Every week, you look at your AI interactions and ask: What worked well? What didn't work? What patterns are emerging? How can we improve our approach?"

She showed Alex a simple framework:

> **WHAT WORKED**: Identify successful AI interactions and understand why
>
> **WHAT DIDN'T**: Analyze failures and extract lessons
>
> **WHAT'S CHANGING**: Notice new patterns or challenges emerging
>
> **WHAT'S NEXT**: Plan one small improvement for the coming week

"The genius of this approach," Maya said, "is that you're constantly getting better at AI without getting distracted by every new feature or model that gets released."

Alex nodded. "So instead of chasing GPT-25, we should focus on getting better results from the AI capabilities we're already using?"

"Right. And here's the interesting part; Ford has found that teams practicing continuous refinement often get better results from older AI models than competitors get from the latest versions."

"How is that possible?"

"Because they understand how to communicate effectively, maintain proper oversight, and apply human judgment. Those skills matter more than which specific AI model you're using."

Maya pulled out her phone. "Let me show you something Ford demonstrated in one of his keynotes."

She showed Alex a side-by-side comparison: the same business task handled by someone using the latest AI model with basic prompts versus someone using an older model with refined communication techniques.

"The refined approach with the older model produced better, more actionable results," Maya said. "Ford's point is that mastery beats novelty every time."

"That's reassuring. But what about staying competitive? If everyone else is upgrading..."

"Ford addresses this directly. He says the competitive advantage isn't having the newest AI, it's having the best AI judgment. While competitors are learning new tools, you're getting better at thinking with AI."

Alex was starting to understand. "So continuous refinement isn't about the technology. It's about the process."

"Exactly. Ford has watched companies across every industry make the same mistake: they upgrade their tools constantly but never upgrade their approach. They're always beginners with new technology instead of experts with established technology."

"And the companies that focus on refinement?"

"They develop compound advantages. Better prompting leads to better results. Better results lead to more confidence. More confidence leads to more experimentation. More experimentation leads to better insights. It's a cycle that builds on itself."

Maya closed her notebook. "Ford shared a story about ancient alchemists spending decades perfecting their processes with the same basic equipment while others chased rumors of magical new ingredients. Guess who actually achieved transformation?"

Alex smiled. "The ones who perfected their process."

"Right. Ford says the same principle applies to AI. Master the fundamentals, refine your approach continuously, and you'll outperform people with fancier tools who haven't mastered the basics."

The next morning, Alex sent a reply to David Kim at Meridian Corp:

"Thanks for reaching out! We're focused on continuously refining our AI approach rather than chasing the latest models. Our results come from better communication with AI tools, not necessarily newer tools. Happy to share insights if you're interested in sustainable AI improvement over cutting-edge upgrades."

David's response came back quickly: *"Interesting approach. We've been upgrading constantly but not seeing the consistent results we expected. Would love to hear more about your refinement process."*

Alex realized something important. While competitors were focused on having the latest AI capabilities, Alex's team was developing something more valuable: the ability to get consistently better results from any AI technology.

That was a competitive advantage that couldn't be easily copied or disrupted by the next AI release.

The Alchemist's Insight

Continuous refinement beats constant upgrading. While competitors chase new AI features, focus on getting better at using the AI capabilities you already have. Mastery of fundamentals always outperforms novelty without substance.

AI won't replace humans, but it will replace humans not using AI. The greatest fear isn't adopting AI too soon, it's waiting too long and falling behind.

Your Next Step

Before adopting any new AI tool, run it through Ford's Refinement Filter. If it doesn't improve results you already value, it's just a distraction. Focus on upgrading your approach before upgrading your tools.

Master the process, and the tools become irrelevant. Chase the tools, and you'll never master the process.

PART III: PRACTICAL APPLICATIONS AND LEADERSHIP

Chapter 11: The AI-Driven Organization

Turning Skeptical Teams Into AI Champions

After months of AI success, Alex thought the hardest part was behind them. They were wrong. The emergency Slack message from Marcus hit at 4:47 PM on a Friday:

"I'm done with AI. I just spent 3 hours fixing a proposal ChatGPT completely botched. Client called it 'impressively wrong.' Need to talk Monday."

Alex stared at the screen, feeling that familiar knot in the stomach. Marcus had been their biggest AI enthusiast just three months ago, the guy who evangelized ChatGPT to anyone who'd listen. If Marcus was giving up, how many others were quietly struggling?

The weekend filled with frustrated Slack threads and half-finished AI projects.

By Monday morning, Alex had called an emergency team meeting. The quarterly regional manager gathering wasn't scheduled for another two weeks, but this couldn't wait.

The conference room had a different energy than usual. Less curiosity, more exhaustion.

"Before we dive into Q3 performance," Alex began, deciding to address it head-on, "I want to talk about what happened with Marcus's proposal. Because I think a lot of us are hitting the same wall."

Marcus shifted uncomfortably. "Look, I probably just screwed up the prompt or something."

"Actually," Jennifer from customer service spoke up, "I've been wondering if I'm the only one who feels like AI is more work than it's worth sometimes."

Heads nodded around the room. Rachel from the Northeast region, who'd been skeptical from day one, looked almost vindicated.

Alex took a breath. "Can I tell you something? Six months ago, I bombed a huge client presentation because I trusted AI too much. The Morrison project. AI gave me confident, comprehensive analysis that was completely wrong for the client's business. I looked like an idiot."

The room went quiet. Marcus looked up.

"So what changed?" Jennifer asked. "Because you've obviously figured something out. Your team's customer satisfaction is up 35%, and you're not drowning in work."

"I learned something from Maya that saved my career," Alex said. "We've been thinking about AI adoption wrong. We keep trying to make everyone use AI the same way. But that's not how it works."

Alex pulled up Marcus's failed proposal on the screen.

"Marcus, walk me through what happened."

Marcus explained how he'd used ChatGPT to generate a comprehensive sales strategy, complete with market analysis and competitive positioning. "It looked professional. But the client said it felt generic and missed their industry completely."

"Jennifer," Alex turned to customer service, "you said AI sometimes works brilliantly for you. When?"

"When I use it for the routine stuff," Jennifer replied. "Password resets, basic product questions, appointment scheduling. It's fast and accurate. But I'd never let it handle a frustrated customer or a complex problem. That's still all human."

Alex nodded. "So Marcus is using AI to create something new and strategic. Jennifer's using AI to handle repetitive tasks. Totally different approaches."

"And I barely use it at all," Rachel added from across the table. "Because every time someone shows me an AI success story, I think of ten ways it could go wrong."

"Exactly," Alex said. "And here's what Maya taught me: we need all three of you."

The room looked skeptical.

"Think about it," Alex continued. "Marcus, you find new AI capabilities before anyone else. Three months ago, you discovered that voice prompting trick that everyone now uses. That's valuable."

Marcus looked slightly less defeated.

"Jennifer, you figured out exactly when AI adds value and when it doesn't. You've probably saved hundreds of hours by automating routine customer service without sacrificing quality on complex issues."

Jennifer nodded.

"And Rachel," Alex turned to their most vocal skeptic, "you're the one who asked if we'd checked data privacy before using AI to analyze customer information. You probably saved us from a lawsuit."

Rachel's expression shifted from defensive to curious.

"Maya calls this the AI-Ready Team approach," Alex explained. "Not everyone needs to be an AI expert. We need different archetypes working together."

Alex sketched five categories on the whiteboard:

The Skeptic - Questions AI adoption, identifies risks, ensures responsible use

Explorer - Discovers new AI capabilities, tests emerging tools

Integrator - Implements AI practically to solve real problems

Strategist - Aligns AI with business objectives, makes decisions

Alchemist - Transforms AI into competitive advantage (the goal we're all working toward)

"Marcus, you're an Explorer. That's not a weakness, it's essential. Three months ago, you discovered that voice prompting trick that everyone now uses.

"Jennifer, you're an Integrator. You figured out exactly when AI adds value and when it doesn't. You've saved hundreds of hours by automating routine customer service without sacrificing quality on complex issues. Jennifer nodded.

"And Rachel," Alex turned to their most vocal skeptic, "you're The Skeptic who asked if we'd checked data privacy before using AI to analyze customer information. You probably saved us from a lawsuit. Skeptics keep us from making expensive mistakes."

Rachel's expression shifted from defensive to curious.

"What about Strategist and Alchemist?" James asked.

"Strategists are the people who make sure our AI efforts actually serve business goals," Alex explained. "They connect AI capabilities to real outcomes. And Alchemists..." Alex paused, "Alchemists are what we're all working toward becoming. People who can consistently transform AI confusion into competitive advantage. That's the journey we're on."

"So if I want to use AI for a client proposal," Marcus worked through it, "I should run it by an Integrator like Jennifer to see if it's actually solving a problem, and a Skeptic like Rachel to check if I'm missing risks?"

"Exactly," Alex said. "Explorers discover capabilities. Integrators implement them practically. Skeptics catch what could go wrong. Strategists ensure it aligns with our goals. And when all four work together, that's when you start operating like an Alchemist."

"This actually makes sense," Rachel admitted. "I thought you were going to tell me I need to become an AI enthusiast. But you're saying my skepticism is part of the system?"

"Your skepticism is essential to the system," Alex corrected. "Without Skeptics, the Explorers would get us all fired."

The room laughed, tension finally breaking.

"I have a question," said James from the West Coast region. "What happens when new AI capabilities emerge? Do we retrain everyone to be Alchemists?"

"That's the beautiful part," Alex replied, remembering Maya's guidance. "We don't force everyone into one archetype. When new AI models or capabilities get released, our Explorers will test them first. Our Integrators will figure out practical applications. Our Skeptics will identify risks. Our Strategists will align them with business goals. The archetypes work together, and over time, everyone develops more Alchemist capabilities."

"So becoming an Alchemist isn't about changing who you are," Jennifer said slowly. "It's about understanding how your natural approach contributes to the bigger picture?"

"Exactly," Alex said. "Alchemists aren't a separate type. They're people who've learned to leverage all five perspectives, whether in themselves or through their team."

Alex pulled up a comparison chart. "While our competitors chase the latest AI features, we've been getting consistently better at AI collaboration. Our net prompter scores are up. customer satisfaction is up 35%, operational efficiency has improved 28%, and most importantly, our team confidence with AI has increased 200%."

"But here's what really matters," Alex added. "Our clients don't keep coming back because we use impressive AI tools. They keep coming back because they trust our judgment about what the AI analysis actually means."

Jennifer raised her hand. "Can I say something? I've been using AI tools for months, and they've made me more productive. But the clients who value my work the most aren't impressed by my AI-generated reports. They're impressed because I explain what the reports mean for their specific situation."

"That's it exactly," Alex said. "AI can help us be faster and smarter, but it can't help us be more trusted. That's still all human."

After the meeting, James approached Alex privately.

"Can I ask you something?" he said. "How do you know if you're building an AI-driven culture versus just implementing AI tools?"

Alex thought about Ford's distinction that Maya had shared. "AI tools give you temporary advantages. AI culture gives you permanent adaptability. When new capabilities emerge, teams with AI culture integrate them naturally. Teams with just AI tools start over every time."

James nodded slowly. "That makes sense. We've been upgrading our AI stack every quarter but not seeing cumulative improvements."

"Maya shared this insight from Ford Saeks," Alex said. "He calls it 'building learning systems, not just technology systems.' The goal isn't having the best AI. It's having the best AI judgment."

"Would you be willing to share Maya's contact information?" James asked. "It sounds like we need to completely rethink our approach."

As Alex wrote down Maya's information, a realization hit. Six months ago, Alex had been desperately seeking AI expertise. Now, competitors were seeking Alex's guidance.

The transformation wasn't just about using AI better. It was about building organizational wisdom that could adapt to any technological change.

That evening, Alex received another email from David Kim at Meridian Corp: "Update: Our latest AI implementation isn't producing the results we expected. Would love to learn more about your team approach."

Alex smiled. While competitors chased shiny new AI models, Alex's team had been building something more valuable: the ability to turn any AI capability into sustainable competitive advantage by understanding how different people work with AI in different ways.

They hadn't just implemented AI tools. They'd built a culture where Explorers, Integrators, Skeptics, and Strategists worked together, with everyone developing their path toward Alchemist-level thinking.

That transformation, from individual archetypes to collaborative alchemy, was the real competitive advantage. And it couldn't be copied by simply buying the latest AI subscription.

The Alchemist's Insight

AI-driven organizations don't require everyone to become Alchemists overnight. They require understanding and respecting five different AI archetypes.

Skeptics question and protect.

Explorers discover possibilities.

Integrators solve practical problems.

Strategists align with business goals.

Alchemists transform capabilities into competitive advantage.

When these five archetypes work together through clear standards and mutual respect, the entire organization develops Alchemist-level capabilities. Success co.

Your Next Step

Discover your AI archetype. Are you The Skeptic who questions and protects? The Explorer who discovers and tests? The Integrator who implements and solves? The Strategist who aligns and decides? Understanding your natural archetype is the first step toward developing Alchemist-level capabilities.

**Find out your AI Archetype. Take the free
AI Readiness Compass™ Assessment at**

AIQuiz.ProfitRichResults.com.

.

| The **Skeptic** | Explorer | Integrator | Strategist | Alchemist |

Chapter 12: Leadership in the Age of AI

Making Decisions That Matter at the Speed of Change

The crisis hit on a Thursday morning at 7:23 AM.

Alex's phone buzzed with an urgent text from Sarah, the CEO: "Major supplier issue. Emergency leadership meeting in 20 minutes. Need options fast."

Six months ago, this kind of emergency would have meant panic, scrambling for data, and walking into the meeting with half-formed ideas. Today felt different. Not because the crisis was less serious, but because Alex had tools that could match the speed of the problem.

The difference wasn't just technology. It was mindset. A calm confidence born from experience applying AI with judgment instead of reacting to it.

By 7:45 AM, Alex was in the conference room with the executive team, laptops open, trying to piece together what had happened.

Their primary supplier had just announced a 40% price increase effective immediately due to "supply chain disruptions and market volatility." This affected 60% of their product line across all their locations.

"How quickly can we assess our options?" Sarah asked.

Alex grabbed a phone and headed to a quiet corner. "Give me fifteen minutes. I'll have preliminary analysis and strategic options."

Alex started with voice commands to Claude while walking back to the office. "Analyze current supplier pricing trends in our industry. I need to understand if this 40% increase is market-wide or specific to our supplier. Focus on alternative supplier pricing and availability."

Back at the desk, Alex opened the laptop and started typing prompts to ChatGPT while the voice analysis was processing.

While typing prompts to ChatGPT: "Help me model the financial impact of a 40% cost increase on 60% of our product line. Include scenarios for absorbing the cost, passing it to our locations, and finding alternative suppliers. Present this for executive decision-making."

While uploading supplier contract images to AI vision tools: "Review these contract terms and identify our options for challenging this price increase or negotiating alternatives."

Simultaneously, Alex used Perplexity's web browsing capability to research the supplier's recent financial performance and competitive landscape.

Fifteen minutes later, Alex had comprehensive analysis that would have taken a team of analysts hours to produce.

"Here's what we know," Alex said, sharing the screen. "The price increase is not market-wide, two of our supplier's competitors are maintaining current pricing. Our contracts include a clause limiting increases to 15% annually. There are three alternative suppliers who could handle 80% of our volume within sixty days."

"Recommendations?" Sarah asked.

"Three options," Alex continued. "Challenge the increase based on contract terms, negotiate a phased increase over six months, or transition to alternative suppliers for our highest-volume products."

The CFO leaned forward. "How confident are you in this analysis?"

"Very confident in the data, moderately confident in the strategic implications," Alex replied. "AI gave me comprehensive information quickly, but I'd want to verify the alternative supplier capacity claims and double-check the contract interpretation before making final recommendations."

Sarah smiled. "That's exactly what I want to hear. Fast insights with clear confidence levels."

The CFO nodded. "Most people would either oversell their certainty or undersell their preparation. This is leadership."

The team made a decision within thirty minutes: challenge the immediate increase, negotiate a phased approach, and begin qualification of alternative suppliers as backup. Alex used AI to draft the initial supplier response, legal strategy memo, and alternative supplier outreach templates.

By noon, the crisis was being managed strategically instead of reactively.

After the meeting, Sarah pulled Alex aside.

"That was impressive. How did you analyze so many variables so quickly while maintaining confidence in your recommendations?"

"What Maya taught me about AI leadership," Alex said. "The goal isn't making faster decisions, it's making smarter decisions faster."

"What's the difference?"

Alex thought about Maya's lessons. "Traditional leadership says gather all available information, then decide. AI leadership says gather the right information quickly, apply human judgment to interpret it, then decide with clear confidence levels about what you know and what you don't."

"So you're not just using AI to make decisions. You're using AI to make better decisions while being honest about uncertainty."

"Exactly. The biggest mistake leaders make with AI is either moving too fast without judgment or moving too slow because they want perfect information. The sweet spot is moving fast with good judgment and clear confidence levels."

Over the following weeks, Alex refined the approach with other leadership challenges. When evaluating a potential acquisition, AI helped analyze financial data and market positioning quickly, but human judgment determined cultural fit and strategic alignment. When planning the annual conference, AI researched venues and logistics efficiently, but human insight shaped the program content and speaker selection.

The pattern was consistent: AI accelerated information gathering and initial analysis, human leadership provided context and judgment, and decisions were made with clear understanding of confidence levels.

"There's something else Ford emphasizes," Alex told the team during a leadership development session. "AI changes the speed of business, but it doesn't change the fundamentals of good leadership."

"What do you mean?" asked David, the operations manager.

"Trust, communication, vision, empathy, and these don't become less important because we have better analytical tools. If anything, they become more important because the technology amplifies everything we do, including our leadership mistakes."

Alex shared an example from the previous month. A location manager had been struggling with staff turnover, and AI analysis had identified several potential factors: compensation below market rate, scheduling conflicts, local competition offering better benefits, and training gaps in the onboarding process.

"Everything AI found was technically accurate," Alex explained. "But when I called the manager to discuss the data, I heard something AI couldn't measure: frustration, defensiveness, and a communication style that immediately put people on edge. Within five minutes, I knew the real issue wasn't any of the factors AI had identified."

The real problem, discovered through that simple phone conversation, was that the manager had poor communication skills that were creating a toxic work environment. No amount of data analysis would have revealed that.

"AI can identify patterns in data," Alex explained, "but it takes human leadership to understand the human story behind the patterns."

"So how do we balance AI insights with human intuition?" asked Rachel.

"Ford's philosophy is simple," Alex replied. "AI gathers information faster than we ever could and finds patterns we might miss. But humans provide the context, make the judgment calls, and handle the relationships. It's not about choosing one or the other, it's about orchestrating both."

When people and AI work in sync, decisions feel smarter and the culture gets stronger.

"The goal," Alex continued, "isn't becoming a more efficient leader. It's becoming a more effective leader who happens to have access to artificial intelligence."

"What's the difference?" James asked.

"Efficient leaders make faster decisions. Effective leaders make better decisions that create lasting value. AI helps with both, but human judgment determines which matters most in any given situation."

The room was quiet as people absorbed the distinction.

"Here's what I've noticed," Alex said. "As AI handles more routine cognitive tasks, human leadership skills become more valuable, not less."

"Like what?"

"Inspiring vision that AI can't generate. Building trust that algorithms can't create. Making judgment calls when data is ambiguous. Managing change when people are scared. Connecting strategy to human values in ways that resonate emotionally."

Alex paused. The lesson was clear. AI does not change who the leader is, it reveals it.

Alex thought about the supplier crisis from earlier. "AI gave us the information we needed to respond quickly, but human leadership decided how to communicate with our site managers, maintain relationships during uncertainty, and balance short-term crisis management with long-term strategic positioning."

"So AI makes us better leaders by handling the analytical work, freeing us to focus on the uniquely human aspects of leadership?" Rachel asked.

"Exactly. The future belongs to leaders who can orchestrate human-AI collaboration, not leaders who can compete with AI or leaders who can hide from it." The work is not to outthink the machine, the work is to help people think better with it.

After the session, Sarah approached Alex.

"I'm curious," she said. "How has your own leadership style changed since learning to work with AI?"

Alex considered the question. "I'm more confident making decisions with incomplete information because I know I can gather additional insights quickly if needed. I'm more focused on asking the right questions because AI can help me find answers faster. And I'm more intentional about the human aspects of leadership because I see how valuable they become when AI handles everything else."

"That sounds like exactly what we need in this environment," Sarah said. "Would you be interested in presenting this approach to the board? They're struggling with AI strategy at the governance level."

As Alex agreed to the board presentation, a realization hit: six months ago, Alex had been learning about AI. Now, Alex was teaching others how to lead with AI. The career restart that had felt so risky eighteen months ago was finally paying off—not just in job security, but in genuine expertise that others valued

The transformation wasn't just about using better tools. It was about developing better leadership judgment in a world where artificial intelligence amplified every decision.

The Alchemist's Insight

AI doesn't replace leadership fundamentals, it amplifies them. The leaders who thrive with AI understand that artificial intelligence handles information while humans handle wisdom. The technology accelerates what you can know, but only human judgment determines what you should do about what you know.

Leadership presence becomes more valuable, not less, when AI handles routine cognitive tasks. Your voice, your judgment, and your integrity are the multipliers that matter.

Your Next Step

For individual leaders: Identify the uniquely human aspects of your leadership role: vision, trust-building, judgment under uncertainty, change management, emotional connection. These are the areas where you add irreplaceable value. Use AI to handle information gathering and analysis, but reserve these human leadership functions for yourself.

Reflect on your last crisis: Could you respond as quickly as Alex did to the supplier emergency? Are you set up to gather intelligence fast while maintaining the human judgment that turns information into wise decisions?

Assess your leadership AI readiness. In Chapter 11, you discovered the five AI archetypes (Skeptic, Explorer, Integrator, Strategist, Alchemist). Now it's time to identify which archetype defines your leadership approach and how prepared your team is for AI-driven decision-making.

Take the free on-demand AI Readiness Compass™ Assessment

at **AIQuiz.ProfitRichResults.com**

For Organizations: Leadership in the age of AI isn't about becoming more efficient at using tools. It's about becoming more effective by knowing when to move fast (AI-powered analysis) and when to slow down (human interpretation, relationship management, and judgment calls that data can't make).

Want to see how Ford Saeks helps leadership teams master AI-human collaboration? Watch his keynote speaker video and explore his executive training programs at **ProfitRichResults.com**.

Chapter 13: Customer Experience Alchemy

How to Scale Personal Touch With Artificial Intelligence

The customer complaint that changed everything arrived on a Wednesday afternoon.

"I've been a loyal customer for three years," the email began, "but your new automated system makes me feel like just another number. I spent twenty minutes trying to explain my problem to a chatbot that kept giving me irrelevant solutions. When I finally reached a human, they had no context about my previous interactions and made me start over completely. I'm considering switching to a competitor who still treats customers like people."

Alex stared at the message, feeling a slow knot tighten in the stomach. It wasn't the first complaint since the AI rollout, but it was the first that felt like a mirror. The Morrison disaster had been about trusting AI too much with analysis. Had Alex now made the opposite mistake, automating customer interactions without preserving what made them human?

The company's AI system was a technical win. Response times had dropped from eight minutes to under two. Routine questions were handled faster than ever. On paper, everything looked golden. But behind the numbers, something human was slipping away.

Alex scrolled through recent survey data. Customer satisfaction up, but loyalty down. The metrics didn't lie: people were getting their answers faster but caring less about who gave them.

Alex called Maya immediately.

"We have a problem," Alex said, voice tight. "Our AI is working too well. Customers feel like they're talking to machines instead of people.

""Ah," Maya replied, calm as always. "You've just met what Ford calls the Customer Experience Paradox—the moment efficiency starts costing empathy. The better AI gets at handling interactions, the more customers crave authentic human connection."

"So what's the solution?" Alex asked. "Turn it off? Go back to human-only service?"

Maya chuckled softly. "If we went backward every time technology outpaced us, we'd still be sending invoices by fax."

She leaned forward. "Ford's answer isn't less AI—it's better AI. He calls it **Customer Experience Alchemy:** using artificial intelligence to scale human touch, not erase it.""

"How does that work?"

"Can you meet tomorrow? This is easier to show than explain."

The next morning, Maya met Alex at their usual coffee shop. The smell of espresso mixed with quiet tension. Alex had barely slept.

"Okay," Maya said, opening her laptop. "Let's start by looking at your customer journey map."

She turned the screen around. "Right now, your system handles initial contact, routes inquiries, and attempts to solve the issue. When it fails, it hands off to a human. So customers experience: robot first, human backup."

Alex groaned. "Exactly. And when the human finally answers, they have no context. The customer has to re-explain everything."

"Ford's model flips that," Maya said, drawing a quick diagram on her notepad. "Human first, AI support. Customers should always feel like they're talking to a person who just happens to have superpowers."

Alex looked skeptical. "Superpowers?"

"Think of it this way," Maya continued. "AI shouldn't replace empathy, it should equip it. Behind the scenes, AI can instantly gather customer history, predict their frustration level, and suggest solutions. But the voice that delivers it must always feel human."

Maya opened a sample case study. "One of Ford's clients implemented AI that analyzed tone and emotion during live calls. The AI quietly coached the rep, suggesting calmer phrasing when tension rose. The customer never knew AI was involved, but they left the call feeling understood."

Alex leaned forward. "So the AI wasn't the star, it was the stagehand."

"Exactly. Ford calls that Invisible AI. Ttechnology that enhances humans without stealing the spotlight. When AI disappears, empathy reappears

"Ford created what he calls the **Personal Touch Protocol**," Maya said, flipping through her notes. "It's a four-part framework that uses AI to strengthen, not weaken, connection."

She listed them one by one:

PREPARATION ALCHEMY: AI gathers context so humans can focus on connection.

PERSONALIZATION ALCHEMY: AI identifies preferences so humans can customize their approach.

ANTICIPATION ALCHEMY: AI predicts needs so humans can proactively help.

FOLLOW-UP ALCHEMY: AI tracks outcomes so humans can maintain relationships.

"Each layer turns data into empathy," Maya said. "The goal isn't speed... it's significance."

Alex nodded slowly. "Show me what this would look like in real life."

"Let's replay your angry customer's call," Maya said. She pulled up a blank notepad and acted it out.

"Customer calls with a problem. Before the rep even says hello, AI has already pulled up their full history, tone markers, and communication preferences. The rep greets them: 'Hi John, thank you for being with us for three years. I see your last two orders had shipping issues. Let's make this one right.'"

Alex smiled. "They feel seen instantly."

"Right," Maya said. "The AI didn't speak. It prepared the human to connect better. Ford always says the best AI experiences are invisible to customers but transformational for employees."

Alex frowned. "But we've got sixty locations. Some of our people barely use email, others are tech wizards. How do we make this consistent?"

Maya grinned. "Ford thought of that too. He calls it AI Training Wheels. It's a layered learning system."

She explained. New reps get full AI support, including suggested phrasing, live emotion analysis, and guided empathy prompts.

Intermediate reps get summaries and recommendations Experienced reps only receive alerts for unusual or high-stakes interactions.

"AI grows with your people instead of overwhelming them," Maya said. "As they develop skill, the AI quietly steps back."

Alex took a deep breath. "So we're not automating empathy. We're scaling it."

Within days, Alex assembled the service team.

"I know some of you think AI is stealing your humanity," Alex began. "But what if it could help you be more human? What if instead of searching for customer info, everything you need appeared the second someone called?"

Jessica from customer service raised a hand. "So we'd sound more prepared without sounding scripted?"

"Exactly," Alex said. "You'll be in control. The AI just makes sure you never walk into a conversation blind."

Within a week, the new approach was live.

The shift was immediate. Reps sounded confident. Customers relaxed faster. Complaints turned into compliments.

Two weeks later, Alex received another email from the same customer who had nearly left. "I don't know what changed, but my recent experience was outstanding. Your rep knew exactly what I'd been through and fixed it without making me repeat everything. It felt personal again. Thank you."

Alex reread it twice, grinning. For the first time in months, the metrics matched the mission. Efficiency and empathy were finally on the same side.

The Alchemist's Insight

The best AI implementations are invisible to customers but transformational for employees. Use AI to make human interactions more personal, not less. Customer Experience Alchemy means scaling personal touch with artificial intelligence rather than replacing personal touch with artificial efficiency.

Customers want to feel heard by humans, not understood by algorithms.

Your Next Step

Audit your customer-facing AI implementations: Are customers interacting with AI or with AI-enhanced humans? Redesign your approach so AI works behind the scenes to give your people superpowers rather than replacing people with systems.

Make AI invisible to customers and invaluable to employees.

Chapter 14: Data and Decision Alchemy

Turning Information Overload Into Competitive Intelligence

Alex actually smiled walking into the monthly performance review meeting.

Eighteen months ago, these sessions meant three days of preparation, mountains of spreadsheets, and charts that never told the real story. Today, Alex had generated the entire quarterly performance dashboard in fifteen minutes over morning coffee, and it was better than anything the old process had produced.

Alex opened the laptop and projected the analysis. Clean, clear, comprehensive. This one generated in fifteen minutes instead of fifteen hours.

"Before we dive into the numbers," Alex said to the executive team, "I want to show you something interesting about our competitive position."

Alex clicked to a chart showing market share trends across their twelve primary markets. "Three months ago, we were losing ground to competitors in Denver and Phoenix. Today, we're gaining market share in both cities."

Sarah leaned forward, curiosity clear in her voice. "What changed?"'she asked, already sensing the story behind the numbers.

"We started practicing what Ford Saeks calls '**Data and Decision Alchemy**'—turning our information overload into competitive intelligence."

Alex explained the transformation. "We were drowning in data but starving for insights. Sales reports, customer feedback, performance metrics, competitor analysis, market trends, and we had more information than we could process, but we weren't learning anything useful from it."

"What was the problem?" asked the CFO.

"We were collecting data instead of asking questions. Ford's approach flips that around. Instead of gathering data and hoping to find insights, you start with the decisions you need to make and then use AI to find the specific information that informs those decisions."

"Let me show you how this worked for our Denver and Phoenix turnaround," Alex continued, clicking to the next slide.

"The decision we needed to make: Should we invest additional support in struggling locations or reallocate resources to high-performing ones? We started there, not with the data."

"Next, we identified what information would actually change our decision. We needed to understand why some locations struggled while others thrived in similar markets."

"Instead of manually analyzing sixty locations worth of performance data, we used AI to identify patterns across sales trends, customer feedback, operational metrics, local market conditions, and competitive positioning."

Alex paused for effect. "AI found something we'd completely missed: the struggling locations weren't performing poorly across all metrics. They were excelling at customer retention but failing at customer acquisition."

"Here's where human wisdom came in," Alex continued. "AI couldn't interpret what this pattern meant. But our understanding of local market dynamics told us these locations had great service but poor visibility in their markets."

"So we invested in targeted marketing support for customer acquisition while maintaining their existing service strengths. Results: 23% increase in new customers within ninety days."

The room was impressed, but Sarah asked the key question: "How is this different from what we were doing before? We always analyzed performance data."

Alex smiled, anticipating this question. "Before, we were using data to confirm what we already thought. Now, we're using data to discover what we didn't know we didn't know."

Alex shared another example. "AI analysis revealed that our highest-performing locations all shared one characteristic that wasn't in our success metrics: they had unusually high repeat purchase rates for specific product categories."

"We would never have looked for that pattern manually. But once AI identified it, we could investigate why those products drove loyalty in those markets and replicate the approach elsewhere."

"That's impressive," said the CFO. "But how do you avoid the information overload problem? AI can analyze everything, but that doesn't mean it should."

"Ford taught Maya that the key is setting what he calls 'Intelligence Boundaries,'" Alex replied. "You define what decisions you're trying to make and only analyze data that affects those specific decisions."

Alex pulled up a simple document. "For strategic decisions like market expansion, we focus on competitive intelligence and market trends. For operational decisions like location support, we analyze performance patterns and local challenges. For tactical decisions like marketing campaigns, we look at customer response data and seasonal patterns."

"Maya taught me Ford's distinction," Alex continued. "Data is what happened. Intelligence is what you should do about what happened."

"So AI helps you turn historical data into future action?" Sarah asked.

"Exactly. But there's one more piece Ford emphasizes, competitive intelligence."

Alex clicked to a new section. "AI can browse competitor websites, analyze their public communications, track their job postings, and monitor their customer feedback in real-time. But the competitive advantage comes from human interpretation of what this intelligence means for our strategy."

Alex showed an example. "AI noticed that three competitors were hiring customer service staff while we were implementing AI-enhanced service. Human analysis recognized this as an opportunity and their costs are increasing while our efficiency is improving, giving us pricing flexibility they don't have."

"This is the kind of strategic insight that creates sustainable competitive advantage," Alex continued. "AI finds the patterns, human wisdom understands the implications."

The CFO raised his hand. "How do you ensure the quality of AI-generated intelligence? We can't afford to make strategic decisions based on flawed analysis."

"Ford's insight is that different decisions need different levels of verification," Alex replied. "For routine decisions, we trust AI with spot-checking. For significant operational decisions, we verify key assumptions. For major strategic decisions, we cross-reference multiple sources and validate critical claims."

"The rule is simple: the bigger the decision, the more human verification is required. But AI still does the heavy lifting of gathering and initial analysis."

After the meeting, Sarah approached Alex privately.

"I'm curious about something," she said. "How has access to all this AI-generated intelligence changed your decision-making process?"

Alex thought for a moment. "I make decisions with more confidence because I have better information. But I also make decisions more carefully because I understand the difference between having information and having understanding."

"What do you mean?"

"AI can tell me what's happening across all our markets simultaneously. But understanding why it's happening and what to do about it, that's still human work. Ford says the goal isn't to make perfect decisions faster, but to make better decisions with clearer confidence about what we know and what we're assuming."

"That sounds like wisdom that extends beyond AI."

"That's Ford's point. Data and Decision Alchemy isn't really about AI. It's about becoming better at turning information into action. AI simply makes it possible to do this at a scale and speed that wasn't possible before."

A week later, Alex received an interesting call from a competitor.

"We've been trying to replicate your market performance in Denver and Phoenix," the caller said. "Our AI analysis shows you're doing something different with customer acquisition, but we can't figure out what. Any insights you'd be willing to share?"

Alex smiled. While competitors were analyzing data, Alex's team was using AI to understand their customers' evolving needs and market opportunities in real-time.

They weren't just collecting information faster, they were transforming information into competitive intelligence that drove strategic action.

The difference between having data and having wisdom had become their sustainable competitive advantage.

The Alchemist's Insight

Information without wisdom is just expensive noise. The competitive advantage isn't having more data, it's turning data into decisions faster and more accurately than competitors. AI excels at finding patterns you'd never see manually, but only human judgment can determine what those patterns mean for your specific situation and strategy.

Data tells you what happened. Intelligence tells you what to do about what happened.

Your Next Step

Start with decisions, not data. Before your next major analysis project, write down the specific decision you need to make and what information would actually change that decision. Then use AI to gather only that relevant intelligence, and reserve the interpretation and strategic application for human judgment.

The goal isn't perfect information, it's better decisions made faster than your competition. Explore Ford Saeks' customized keynotes and workshops for AI Alchemy at
 ProfitRichResults.com

Chapter 15: Preparing for the Future

What's Coming Next and How to Be Ready for It

One year after meeting Maya, Alex received an invitation that would have seemed impossible eighteen months earlier.

"We'd like you to keynote our annual conference on AI transformation in business," the email read. "Your organization's approach to human-AI collaboration has become a model for sustainable AI adoption across multiple industries."

Alex stared at the screen. A year ago, Alex had been desperately seeking AI expertise. Now, other leaders were seeking Alex's guidance.

The irony wasn't lost: Alex had become an AI expert not by mastering technology, but by mastering the human side of artificial intelligence.

That evening, Alex called Maya to share the news.

"I'm not surprised," Maya said. "You've been practicing what Ford calls '**Future-Ready Leadership**', developing capabilities that adapt to change rather than just reacting to specific technologies."

"But here's what I'm wondering," Alex said. "AI is evolving so fast. Autonomous agents, advanced reasoning, multimodal capabilities, workflow automation so how do I prepare for changes I can't even imagine yet?"

"Perfect question. Ford addresses this constantly in his keynotes. He says the secret to preparing for an unpredictable AI future isn't trying to predict what's coming, it's developing what he calls '**Adaptive Intelligence**.'"

"What's Adaptive Intelligence?"

"The ability to learn, apply, and refine your approach to any new AI capability that emerges. Ford has identified five core competencies that remain valuable regardless of how AI evolves."

Maya opened her notebook to a page titled "Future-Proof Capabilities."

"First, **Strategic Thinking.** AI can process information and identify patterns, but it can't determine what those patterns mean for your specific strategy. That's uniquely human."

"Second, **Relationship Mastery.** As AI handles more routine interactions, the ability to build authentic human connections becomes increasingly valuable. Ford predicts a premium market for human-to-human business relationships."

Alex thought about the keynote invitation. "That makes sense. People invited me to speak because they want human insight about AI, not AI-generated content about AI."

"Exactly. Third, **Judgment Under Uncertainty.** AI provides analysis, but humans make decisions when information is incomplete or conflicting. That skill becomes more important as AI gives us more information to evaluate."

"Fourth, **Creative Problem-Solving.** AI can suggest solutions based on existing patterns, but breakthrough innovations come from human creativity that sees possibilities AI hasn't been trained to recognize."

"And fifth?"

"**Adaptive Learning.** The ability to quickly understand new AI capabilities and integrate them into existing expertise. Ford says this is the meta-skill that makes all other skills more powerful."

Alex was taking notes. "So instead of trying to predict specific AI advances, I should focus on developing these five capabilities?"

"Right. Ford's philosophy is that these competencies will make you more valuable whether future AI can do everything current AI does, plus see, hear, reason, browse the web, and execute complex workflows autonomously, or whether it develops capabilities we can't even imagine yet."

"Can you give me an example?"

Maya pulled up her phone. "Ford recently worked with a manufacturing company preparing for AI that might eventually manage entire production lines autonomously. Instead of trying to predict exactly how that would work, they focused on developing human capabilities that would remain valuable: strategic oversight, quality judgment, relationship management with suppliers and customers, creative problem-solving for unusual situations, and adaptive learning for integrating new AI capabilities as they emerge."

"What happened?"

"When advanced AI production tools became available, they were ready. Not because they'd predicted the specific technology, but because they'd developed the human capabilities needed to work with any production AI effectively."

Alex nodded. "So future-proofing isn't about the technology. It's about the thinking."

"Ford puts it this way: 'Prepare for the future by becoming more valuable to the future.' AI will continue evolving, but the need for strategic human judgment, authentic relationships, and adaptive leadership will only increase."

"What about staying current with AI developments? Should I be tracking every new release and capability?"

"Ford recommends what he calls 'Strategic Awareness, Not Feature Anxiety.' Stay informed about major AI trends that could affect your industry, but don't chase every new feature or model."

Maya showed Alex Ford's approach: "Monthly review of significant AI developments, quarterly assessment of how new capabilities might affect your business, annual strategic planning that incorporates AI evolution, but daily focus on mastering current AI applications."

"The goal is being prepared for change without being paralyzed by change," Maya continued. "Ford says the leaders who thrive in the AI future are those who develop strong fundamentals and apply them consistently, not those who constantly restart with new technologies."

Alex thought about the past year's journey. "You know what's interesting? The frameworks you taught me: purpose over technology, CRAFT communication, and balancing speed with judgment. They've all become more important as AI has gotten more sophisticated, not less."

"That's Ford's point exactly. The principles are timeless because they're about human nature and good decision-making, not about specific technologies."

"So what's my responsibility as a leader preparing my organization for an AI future we can't fully predict?"

Maya smiled. "Ford would say your job is to build a learning organization that can adapt to any AI capability that emerges. Focus on developing people who can think strategically, communicate effectively, build relationships authentically, and learn continuously."

"And the AI part?"

"The AI part becomes easy when you have the human part figured out. AI amplifies everything, including good judgment, bad judgment, clear communication, confused communication, strategic thinking, reactive thinking. The better your human foundations, the better your AI results, regardless of which AI capabilities emerge."

Six months later, Alex was indeed delivering that keynote presentation to a room full of executives facing their own AI transformation challenges.

"The secret to AI leadership," Alex told the audience, "isn't predicting the future of artificial intelligence. It's preparing yourself and your organization to thrive with any AI future that emerges."

Alex shared the story of the journey from AI confusion to confident mastery, emphasizing Ford's frameworks and Maya's guidance.

"The ancient alchemists spent centuries trying to turn lead into gold," Alex concluded. "They never succeeded at creating precious metal, but they succeeded at something more valuable: they developed the scientific method, established the foundations of modern chemistry, and proved that transformation is possible when you understand the underlying principles."

"Today's AI alchemists face a similar challenge. We're not trying to predict which AI capabilities will emerge next. We're developing the timeless principles that help us transform any AI capability into sustainable competitive advantage."

"Whether you're dealing with today's ChatGPT, tomorrow's autonomous agents, or whatever gets invented next year, the principles remain the same: Start with purpose, communicate clearly, balance speed with judgment, build trust through transparency, and refine continuously."

"Master these principles, and you'll be ready for any AI future. Because the future belongs not to those who have the best AI, but to those who can think best with AI."

After the keynote, several audience members approached Alex with questions about implementing Ford's frameworks in their organizations.

One executive asked, "How do you stay confident about the future when AI is changing so rapidly?"

Alex smiled, remembering Maya's final lesson: "Ford taught me that confidence comes from adaptability, not predictability. I can't predict what AI will be capable of in five years, but I'm confident in my ability to evaluate, integrate, and refine whatever capabilities emerge."

"That's the real alchemy," Alex added. "Transforming uncertainty about the future into confidence about your ability to adapt to any future."

Six months ago, Alex had stared at a blinking cursor, terrified of being left behind... again. Now, standing in front of sixty franchise owners who actually understood how to use AI strategically, Alex realized something: the career restart wasn't a desperate pivot. It was preparation for exactly this moment.

Sarah had been right to bet on Alex after all.

After the keynote, several audience members approached Alex with questions about implementing Ford's frameworks in their organizations.

The Alchemist's Insight

Prepare for the future by becoming more valuable to the future. Develop the five future-proof capabilities: Strategic Thinking, Relationship Mastery, Judgment Under Uncertainty, Creative Problem-Solving, and Adaptive Learning. These skills become more valuable as AI becomes more capable.

The future belongs not to those who have the best AI, but to those who can think best with AI.

Your Next Step

Focus on developing adaptive intelligence rather than chasing specific AI capabilities. Identify which of the five future-proof competencies need strengthening in your role:

- ‣ Strategic Thinking
- ‣ Relationship Mastery
- ‣ Judgment Under Uncertainty
- ‣ Creative Problem-Solving
- ‣ and Adaptive Learning.

Use current AI tools to practice these skills. Each AI interaction is an opportunity to strengthen your adaptive intelligence.

Master the principles, and you'll be ready for any AI future that emerges.

Want to build future-ready leadership? Explore Ford Saeks' keynote presentations and training programs at ProfitRichResults.com

CONCLUSION

Your AI Alchemy Roadmap:

Alex's story ends here, but your journey is just beginning.

I created Alex to represent thousands of leaders I've worked with: smart professionals who felt lost when AI entered their world. Alex's transformation from confusion to confidence isn't fiction. It's the same journey I've guided countless executives through using these frameworks.

The pattern is always identical: leaders focus on technology instead of outcomes, chase features instead of mastering fundamentals, and try to compete with AI instead of collaborating with it.

But when they shift their approach, everything changes. AI becomes a powerful ally instead of a confusing obstacle.

Your 30-Day Quick-Start

Week 1: Start with Purpose - Apply the Alchemist's First Rule to three challenges. Ask: What am I achieving? Why does this matter? How will I know success?

Week 2: Master Communication - Practice the CRAFT method and use AI to help you craft better prompts.

Week 3: Balance Speed with Wisdom - Implement the Speed-Check System and start your Weekly Refinement Ritual.

Week 4: Build Trust - Have honest AI conversations with your team, focusing on human amplification.

Beyond: Develop the five future-proof capabilities: Strategic Thinking, Relationship Mastery, Judgment Under Uncertainty, Creative Problem-Solving, and Adaptive Learning.

Your Sustainable Advantage

While competitors chase the latest AI features, you'll develop something more valuable: the ability to think strategically with any AI capability that emerges.

While others implement tools, you'll build culture. While others learn platforms, you'll master principles.

This compounds over time and can't be replicated by upgrading technology alone.

The Human Premium

Here's what I've been predicting in keynotes worldwide: as our world becomes automated, human-to-human connection becomes premium.

The more AI handles routine interactions, the more valuable authentic human interaction becomes. Executives don't invite me to speak for AI information, they want human wisdom about AI integration.

The future belongs to leaders who amplify human purpose with artificial intelligence.

Your Transformation Starts Now

The ancient alchemists never turned lead to gold, but they proved transformation possible through understanding principles.

You face the same opportunity: developing wisdom to thrive with any AI future.

Transform your relationship with AI, and you transform everything: productivity, decision-making, leadership effectiveness, competitive positioning, and confidence about the future.

The goal isn't becoming someone who uses AI. It's becoming someone AI makes better.

Start with one framework. Apply it consistently. Build your adaptive intelligence.

The principles remain constant:

Start with purpose. Communicate clearly. Balance speed with judgment. Build trust through transparency. Refine continuously. Stay human.

The question isn't whether AI will evolve. It will.

The question is whether you'll be ready to evolve with it.

Your alchemy begins with the next decision you make about approaching artificial intelligence.

Choose wisely. Your transformation awaits.

AI Alchemist's Quick-Start Checklist

Steps to Put AI Alchemy into Action Today

❑ Start with Purpose

Before you open an AI tool, ask: What outcome am I trying to achieve? Why does this matter? (Alchemist's First Rule)

❑ Use the CRAFT Method

Context • Role • Audience • Format • Tone.

Every AI interaction starts with a clear, CRAFTed instruction.

❑ Run the Speed-Check System

- Routine tasks → trust AI.
- High-stakes decisions → verify with human judgment.

❑ Challenge the Output

Don't accept the first draft. Ask AI to expand, reframe, or argue against its own answer. This builds stronger, more creative results

❑ Apply the Weekly Refinement Ritual

Take 15 minutes each week to review what worked, what didn't, and how you'll improve your AI collaboration next week.

❑ Practice Cross-Prompting

Don't stay locked into one platform. Test the same prompt in multiple tools (Perplexity, ChatGPT, Claude, Copilot, Gemini, etc.) and compare outputs.

❑ Keep Humans in the Loop

AI should accelerate your thinking, not replace it. Use AI to generate, but rely on human judgment to filter, refine, and decide.

❑ Set Guardrails for Your Organization

- Establish clear AI Acceptable Use Policies (AUP).
- Train your team to use AI responsibly and ethically.
- Keep compliance, privacy, and brand voice in mind.

❑ Invest in Human Skills

Balance your AI learning by developing the skills machines can't replace: empathy, storytelling, strategic judgment, and ethical leadership.

❑ Build a Culture of Conversation

Create a dedicated MSTeams or Slack channel where your team can share wins, experiments, and lessons learned. Champion ongoing dialogue around AI adoption.

❑ Document and Share Wins

Capture screenshots, drafts, or outcomes where AI added real value. Share them with your team to build confidence and momentum.

❑ Stay Adaptive

Remember: today's tools are multimodal and evolving fast. Focus on mastering the principles so you can adapt as capabilities expand.

TIP: *Return to this checklist often. Keep it bookmarked in your copy of AI Alchemy and review it weekly. Mastery isn't about chasing every new tool, it's about practicing timeless principles, refining your approach, and leading your team with confidence.*

Prompting Guide

How to talk with AI so it works like a partner, not a parrot

❑ The CRAFT Framework

Use this structure every time you prompt:

C = Context → What's the background or situation?

R = Role → Who should the AI act as? (coach, analyst, strategist, editor, etc.)

A = Audience → Who is this output for? (executives, customers, students, etc.)

F = Format → What shape do you want the output in? (email, table, outline, script)

T = Tone → How should it sound? (professional, empathetic, playful, concise)

❑ Pro Tips for Alchemist-Level Prompting

Cross-Prompting: Run the same prompt in multiple tools (ChatGPT, Claude, Copilot, Gemini, Perplexity) → compare → refine.

Dialogue, Not Monologue: Always end with: "What else do you need to know? What am I forgetting?" to spark iteration.

Keep Humans in the Loop: Treat AI as a collaborator. You filter, refine, and decide.

Challenge the Output: Ask AI to argue against itself, expand, or generate alternatives.

Refine Weekly: Save your best prompts in a shared doc or Slack channel. Champion small wins with your team.

Example Prompts

1. Leadership Communication

"Context: I'm preparing a leadership update for my executive team. Role: Act as a communication coach. Audience: Senior executives who want clarity and actionable insights. Format: A concise, 3-paragraph email update. Tone: Professional, confident, and strategic.

What else do you need to know before drafting this?"

2. Customer Service Response

"Context: A customer is frustrated about a delayed shipment. Role: Act as a customer experience specialist. Audience: A long-time customer. Format: A 2-paragraph email reply. Tone: Empathetic, apologetic, but solution-focused.

What am I forgetting that would help you write a better reply?"

3. Innovation Brainstorm

"Context: Our franchise team wants to explore new revenue streams. Role: Act as an innovation facilitator. Audience: Franchise owners. Format: A brainstorm list of 10 fresh ideas. Tone: Creative, bold, and business-savvy.

What assumptions should I challenge to get better ideas?"

Remember: Mastery isn't about knowing every AI tool. It's about learning how to ask better questions and refining the answers until they serve your purpose.

About the Author

Ford Saeks: Business Growth Accelerator and AI Transformation Expert

Ford is a Hall of Fame keynote speaker and Business Growth Accelerator who has spent over thirty years helping organizations navigate transformative change. As President and CEO of Prime Concepts Group, Inc., he has guided companies from startups to Fortune 500 enterprises through periods of rapid technological evolution, generating over a billion dollars in client sales through his strategic guidance.

Ford's expertise in AI transformation stems from his unique position at the intersection of technology and human psychology. Unlike consultants who focus purely on technical implementation, Ford specializes in the human side of artificial intelligence. He helps leaders overcome AI overwhelm and build sustainable competitive advantages through human-AI collaboration

As author of *AI Mindshift*, *Accelerate*, and *Superpower,* Ford brings real-world experience from working with franchise systems, corporate teams, and associations worldwide. His approach to AI alchemy, and the principles shared throughout this book, has been refined through thousands of hours helping actual organizations implement AI successfully without losing their humanity.

Ford holds three U.S. patents, has founded over ten companies, and has delivered more than 2,500 presentations globally. His insights have been featured in Fast Company, Wall Street Journal, NBC News, Inc. Magazine, Entrepreneur, Bloomberg Businessweek, and CBS.

When he's not on stage or consulting with organizations, Ford enjoys exploring how ancient wisdom applies to modern challenges. That inspiration drives AI Alchemy's unique approach to technological transformation.

Bring Ford Saeks into Your Organization

Transform Your Team from AI Overwhelm to Competitive Advantage

Ready to help your organization master the art of human-AI collaboration? Ford Saeks delivers dynamic keynote presentations and interactive workshops that give your audience the same frameworks that transformed Alex's journey from confusion to confidence.

Featured Keynote: "AI Alchemy: From Overwhelm to Competitive Edge"

This presentation takes your audience through the exact transformation process outlined in this book. Using compelling storytelling and practical frameworks, Ford shows leaders how to:

- Apply the Alchemist's First Rule to make AI serve human purpose
- Use the CRAFT method for better AI communication across all modalities
- Implement the Speed-Check System for balancing AI efficiency with human judgment
- Build trust during AI transformation using the principles from "Avoiding the Dark Arts"
- Create sustainable competitive advantages through continuous refinement

What Your Audience Will Experience:

- **Immediate Relief**: Understanding that AI mastery isn't about technical expertise, it's about better thinking
- **Practical Frameworks**: Tools they can implement the day after your event

- **Confidence Building**: Clear roadmap for staying ahead of AI evolution without chasing every new feature
- **Human-Centered Approach**: How to use AI while becoming more valuable, not replaceable

Why Organizations Choose Ford:

Proven Track Record: Over 500,000 professionals trained across every industry **Relevant Experience**: Deep understanding of franchise operations, corporate dynamics, and association challenges

Engaging Delivery: Hall of Fame speaking ability that combines entertainment with education

Customized Content: Every presentation tailored to your specific industry and audience needs

Lasting Impact: Audiences leave with immediately actionable strategies, not just inspiration

Watch Ford's Speaker Trailer Video

Unlock the Power of AI... Without the Pitfalls!

AI MASTERCLASS VIDEO TRAINING

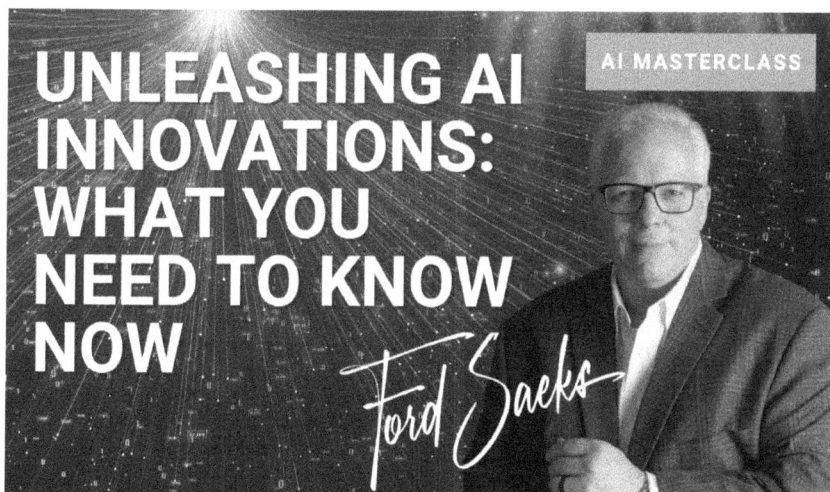

On-demand Video Training that complements this book's frameworks, visit **ProfitRichResults.com/ai-training**.

This program provides practical implementation guidance for individuals and teams across all departments.

Discover how to leverage AI tools effectively and avoid common missteps that can harm your brand. In this exclusive Video Training, Ford Saeks breaks down:

- Essential AI strategies and tactics you and your team can use right now.

- Key mindset shifts for effective AI integration, regardless of your role

- Practical steps to build AI literacy and empower your team, with no theory, just relevant takeaways... guaranteed!

This is the one AI training video to share with your entire team. Get actionable insights to drive innovation, efficiency, and sustainable growth. This is updated regularly, so get instant access right now to the video, companion guide, and bonus templates.

Additional Reading:

AI Mindshift: *How to Unleash the Power of AI, Avoid the Pitfalls, and Keep the Human Touch*

Accelerate: *The Ultimate Guide for Franchisees to Maximize Local Marketing and Boost Sales*

Superpower: *A Superhero's Guide to Leadership, Business, and Life*

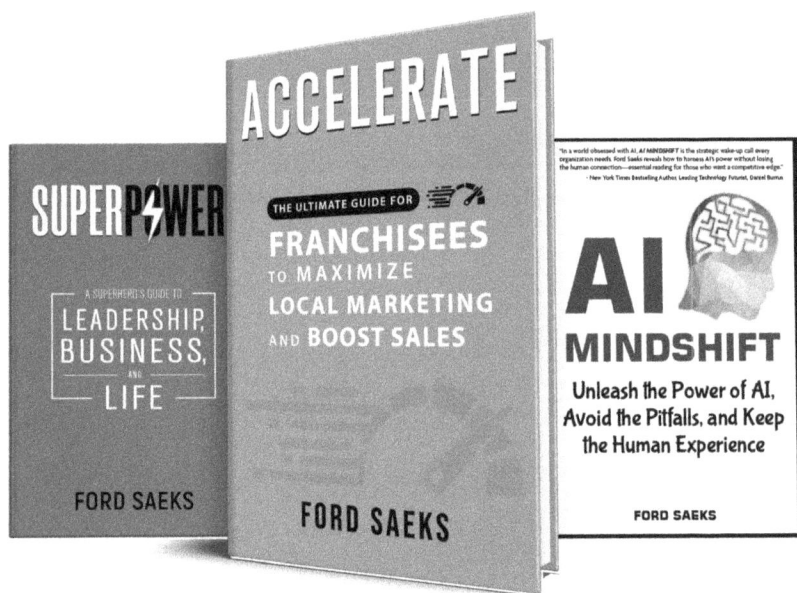

Stay Connected

Speaking Inquiries: Bring Ford's AI Alchemy keynote to your next event

- Email: FordSpeaks@ProfitRichResults.com

- Phone: 316-844-0235

- Website: www.ProfitRichResults.com

- LinkedIn: www.linkedin.com/in/fordsaeks

Additional Resources:

- Watch Ford in action Weekly at 11 : Fordify.tv

- Listen to *The Business Growth Show* podcast on your favorite platform

- Follow ongoing AI insights and updates at ProfitRichResults.com

Your AI Alchemy Starts Now

The principles in this book work immediately, but they compound over time. While your competitors chase the latest AI features, you'll be developing something more valuable: the ability to think strategically with any AI capability that emerges.

Start with one framework. Apply it consistently. Share your learnings. Build your adaptive intelligence.

Remember: *AI won't replace humans, it will replace humans not using AI.*

The question isn't whether to adapt to this technology. The question is whether you want to be in control of how you adapt.

Your transformation from AI overwhelm to competitive advantage begins with the next decision you make about approaching artificial intelligence.

Choose wisely. The future belongs to AI alchemists.

www.ingramcontent.com/pod-product-compliance
Lightning Source LLC
Chambersburg PA
CBHW042121190326
41519CB00031B/7576